Mobile Learning

Perspectives on Practice and Policy

A Volume in Digital Media and Learning

Digital Media and Learning

Danielle Herro and Remi Kalir, Series Editors

Mobile Learning: Perspectives on Practice and Policy (2018)
Edited by Danielle Herro, Sousan Arafeh,
Richard Ling, and Chris Holden

Mobile Learning

Perspectives on Practice and Policy

Edited by

Danielle Herro
Clemson University

Sousan Arafeh
Southern Connecticut State University

Richard Ling
Nanyang Technological University

and

Chris Holden
University of New Mexico

IAP

Information Age Publishing, Inc.
Charlotte, North Carolina • www.infoagepub.com

Library of Congress Cataloging-in-Publication Data

CIP data for this book can be found on the Library of Congress website:
http://www.loc.gov/index.html

Paperback: 978-1-64113-122-3
Hardcover: 978-1-64113-123-0
E-Book: 978-1-64113-124-7

Names: Herro, Danielle, editor. | Arafeh, Sousan, editor.
Title: Mobile learning: perspectives on practice and policy / edited by Danielle Herro and Sousan Arafeh.
Other titles: Mobile learning (Information Age Publishing)
Description: Charlotte, North Carolina: Information Age Publishing, Inc., 2018. | Series: Digital media and learning | Includes bibliographical references.
Identifiers: LCCN 2017050342 (print) | LCCN 2018003494 (ebook) | ISBN 9781641131247 (EBook) | ISBN 9781641131223 (pbk.) | ISBN 9781641131230 (hardcover)
Subjects: LCSH: Mobile communication systems in education—United States. | Internet in education—United States.
Classification: LCC LB1044.84 (ebook) | LCC LB1044.84 .M614 2018 (print) | DDC 371.33—dc23
LC record available at https://lccn.loc.gov/2017050342

Printed and bound by CPI Group (UK) Ltd, Croydon, CR0 4YY

CONTENTS

FOREWORD

Context and Questions

John Traxler
Professor of Digital Learning,
Institute of Education, University of Wolverhampton

Mobile learning, whatever it may be exactly, is spreading fast and changing fast, and certainly moving away from its roots.

Its history to date has been a patchwork of achievements, mistakes, insights, platitudes, and epiphanies as the ground has moved under us. It has moved from a predictable landscape where mobile learning researchers were the innovators and early adopters in their institutes and universities, working with novel, expensive, difficult, fragile, and scarce technologies fueled by curiosity on a trajectory defined by the aspirations of tethered e-learning. It has moved to a much less predictable environment where the technologies had become cheap, easy, robust, and universal, indeed intrusive, and understood, appropriated and exploited by everyone.

Early research, with findings predicated on consistent, stable technological platforms, looked irrelevant as it became more apparent that the short term teaches us little about the long term, the small scale teaches us little about the large scale, and the pedagogic and technological answers do not help with the organizational and cultural questions. So, dear reader, history will not be teaching us much and when it comes to mobile learning, we are on our own, scrutinizing the current texts as carefully as

Mobile Learning: Perspectives on Practice and Policy
pp. vii–xii

possible and trying to discern the message and its meaning. We need these new texts to capture the present before it becomes the irrelevant past. The readers of mobile learning are perpetually challenged to look at the dots, that is, all the accounts of individual researchers, projects, pilots, programs, and interventions, and join them up, into some bigger, broader, more abstract, generalized, conceptualized picture—sadly there are never enough dots, they fade before our eyes as they recede into history and we never get to the *big picture* of mobile learning. So instead we must scrutinize the dots, the individual accounts in the current book, and hope for whatever insights resonate with our own contingent problems and perspectives.

While this may be a slightly apocalyptic portrayal we must also recognize that mobile learning is a recognized and established phenomenon within the various sectors and countries of formal education. We must not only interrogate these texts to understand the role of mobile learning in the context of mobile and connected societies but also within the ecology of education technology. This wider education technology environment has moved on and mobile learning has to compete, complement, contribute to, or coalesce with, for example, ideas about the flipped class-room, massive open online courses, learner analytics, *bring-your-own-device* (BYOD), lecture capture, digital literacy, heutagogy, and content curation, with plagiarism detection, online submission, and each new generation of virtual learning environment as neighbors in the educational technology ecosystem. To some extent, the longer term viability of mobile learning, and the projects and pilots described here, depends on how it is perceived as a component or a competitor or a complement to any of these other trends, vying for resources, capacity and change. Some are conceptual and theoretical, for example heutagogy and the connectivism that underpinned the early massive open online courses, some are operational and pragmatic, for example lecture capture and plagiarism detection, and that makes a difference. Features and attributes that might appeal to innovators and early adopters might not be what appeals to regulators, managers, and rank-and-file faculty; the latter sometimes preferring whatever seems to offer a pragmatic, a more efficient version of the *status quo* while the former seek novelty and surprise, and conceptual transformation. This one of the challenges for innovation in mobile learning.

The flipped classroom was originally an American conception and was a response to various practical and pedagogic problems and challenges (principally those of encouraging autonomous active learning, discussions about BYOD, and increasing pressure on contact time) and a response to a specific infrastructure and technical environment—namely the widespread availability of cheap connectivity and personal, social, and domestic digital technologies. While an underpinning theory of the flipped classroom may have emerged, the idea was nevertheless originally nationally specific and culturally specific, and so the documented experiences come originally from a very specific setting, namely schools in America. The flipped classroom concept should not be interpreted merely as using technology, for example mobile technology, to displace the consumption of content out of the classroom. In fact, the principles of the flipped classroom can be expressed as the following quite profound questions or challenges, in particular, How can we optimize the face-to-face learning experience? What is it that students can only get from face-to-face interaction with lecturers? How can we optimize the ways in which students can learn from each other face-to-face? What is it that students can only get face-to-face with each other? How can we optimize the campus experience? What is it that students can only get by coming on campus? And, how can digital technology, including obviously mobile digital technology, support addressing these challenges? And what does digital technology, including mobile do most effectively or uniquely?

These kinds of questions are being discussed though perhaps not very rigorously. They are linked indirectly to earlier discussions about *personal response systems,* namely discussions about promoting more active and engaged learning in formal education. While its first incarnation of personal response systems was delivered on dedicated devices, the so-called *clickers* or *zappers*—and thus an unsustainable expense—later incarnations use software or apps, such as *Socrative,* that run on learners' own devices and thus align with emergent BYOD policies. Lecture capture, another current trend, is a technical fix to the problem of student attendance, English language proficiency and enhanced revision, but is often implemented on systems, such as *Panopto,* that encourage playback on mobiles, and thus also aligned BYOD policies.

These policies are increasingly visible and the relationship with mobile learning is fairly clear. Furthermore, they seem to dig mobile learning out of the hole of its own making. That is the hole of unsustainability. At a more operational level, however, BYOD policies present several local or tactical problems. First, the technical and infrastructural challenges, secondly the operational aspects and thirdly, the ethical and political issues. Mobile technologies are obviously the devices in question, albeit rather vaguely defined, but mobile learning has largely developed independently of the BYOD debate and discussion.

Content curation, sometimes called orchestration, is the concept of supporting learning with resources from wherever they were best, not defaulting automatically to the treadmill of in-house content creation. If content, and its intellectual property, represent a perceived business asset within their respective universities then curation is the basis for a very different business model. It is kindred to the open learning movement, driven by motivation that is part pedagogic, part ideological, part practical. To work effectively curation requires teachers (or librarians or the designers of learning experiences) to develop skills around the selection and organization of resources. Given that the mobile is increasingly a portal to the resources of the web for an almost universal demographic beyond the classes and institutions that access networked computers, there is a clear and powerful connection between mobile learning and content curation but higher education is still largely committed to in-house resources. Skills and attitudes are the likely barrier to change, though it is perfectly reasonable to argue that in a few cases in a few institutions in-house content encapsulates unique new knowledge, unavailable anywhere external.

The movement to exploit external content, that is the curation of content, is aligned to the movement to exploit user-generated or learner-generated content. This is not necessarily a mobile phenomenon but the mobile empowers a more broadly-based demographic of creators and richer and diverse content, derived from the environment and the context, wherever people could take their mobiles. It implicitly happens every time someone shares their ideas and images on social media (and similarly, curation happens every time someone uses social book-marking) but is only a fringe activity in higher education.

Heutagogy, or self-directed learning, is an emerging pedagogic paradigm, in some ways a rejuvenation of informal lifelong learning and a recognition that in an era where everything changes, everyone needs to continually learn, to continually learn digitally. At the moment however, the connection between mobile learning in higher education and heutagogy during and after higher education is tenuous and receiving little attention.

Digital literacy is the pervasive and comprehensive underpinning of all these various trends and formulations. In its broadest interpretation, digital literacy describes those skills, attitudes, access and competences necessary for individuals, and perhaps communities, to flourish in an increasingly digital world. The genesis of digital literacy was higher education in western Europe with a clearly defined pressure to focus on e-learning, information technology skills, and employability, and a less coherent lobby for the social, culture, expressive, and creative aspects of being digital. A specifically mobile element of digital literacy has yet to be developed in spite of the observation about how pervasive and ubiquitous mobiles are in our societies, and this may illustrate the tension between the educational technologies of the institutions as opposed to the universal technologies of their societies.

A discussion of the MOOC should follow on logically from all the previous topics. The idea of the MOOC was born out of experiences with large open on-line courses in higher education that suggested a new pedagogy, where the numbers and connections would create a new learning paradigm. The story of the MOOC is however not straightforward. In its original formulation, it did indeed hark back to many of the issues just discussed. It grew out of ideas and experiences with large numbers of connected learners, termed *connectivism.* As well as the well-documented Canadian experiments, there were small ones with a specifically free and mobile focus, specifically MobiMOOC. The idea of the MOOC became, however, co-opted by formal institutional perspectives and now has been transformed into a highly interactive rich media experience broadcast by universities on a small number of specialized and dedicated platforms. Some platforms more than others render well on mobile devices—*FutureLearn* in the United Kingdom is a prime example of a *mobile first* MOOC philosophy—but the early idealism of the *wisdom of the crowd* has been replaced by a globally competitive and corporate ethos.

Alongside the issues of the changes, fashions, and trends that constitute the immediate environment of mobile learning we must also recognize elements of stasis and stagnation; whatever the rhetoric, the learning management system or virtual learning environment has become the established repository of institutional content and change is often only represented by periodic replacement of one model or brand by its obvious successor rather than by any more fundamental debate. In this sense virtual learning environment or learning management system have become an established part of the university experience or the university tradition, of which learner analytics, plagiarism detection and online submission are just minor appendages or refinements. Consequently, in this depiction, mobile learning may only mean accessing the virtual learning environment or receiving short message service alerts on a mobile, and innovation is necessarily severely constrained at an institutional level; the enacted version of mobile learning in education may be quite different from the espoused version enunciated by its early fathers. The current contributions break out of this narrow depiction.

Perhaps, as preparation for diving into the contributions, we should all get our bearings and ask ourselves a few questions. Given the novelty, change, potential, interdisciplinarity, diversity, and contexts of mobile learning, it probably requires a much more active and critical readership than some other more established disciplines.

First, given this impressionistic overview of the context of the current book, what does each of us readers think *mobile learning* is, both in theory and in practice, and as something we espouse and something we enact? This is perhaps the recurrent definitions question; how does each of us define mobile learning in our writing and in our work? Is it a component of e-learning mostly in educational institutions or is it a pervasive feature of our societies?

Where is each of us coming from and how has that shaped our perceptions and preferences? Is it from a technical background or an educational background? Or is it from background in research driven by theorizing, speculating, and curiosity or in research coming from a desire to spread learning further and wider?

Actually, those of us who are innovators or early adopters—sadly the digital learning community is still shackled to the lexicon of the *diffusion of innovations*—should be asking ourselves about the extent of our understanding of the values, concerns, and constraints of regular teachers and lecturers and their managers. How can we, the current readers and contributors, connect with them all?

So, now primed with the context and the questions, read on.

PREFACE

When it comes down to it, education has yet to take much notice of "mobile." For most of the last decade, aside from dealing with new crises like active shooter scenarios or questions of who should use what bathroom, educators have largely continued business-as-usual. They have often ignored the potential of mobile technologies for learning, even when pioneering teachers, researchers, and schools tinker at the edges. Meanwhile, mobile technology has assumed a central position in society.

In this context, education's intransigence should not be mistaken for mere luddism—though there's certainly a fair share of that too. Instead, the reluctance reflects how large the challenge of mobile communication is. It takes time and effort to turn a large machine around and schools do not turn on dimes, which can be both good and bad. However, the social changes inspired by mobile technological developments *are* extensive. Mobile learning moves us beyond simple questions about how instruction is delivered or whether it should be compartmentalized within domains or between pedagogy and policy. Mobile learning has been elusive. In part, because it is still a missed opportunity; however, it is also because so many good ideas about mobile learning have not gained enough traction to thrive. The "no-easy-answers" that live in the wile tempt, but too often disappoint. Through this book we seek to offer touchstones in the continued quest for more extensive awareness and adoption of mobile learning.

There is one seemingly easy answer to the challenges of mobile learning to which we bear witness: its outright ban. Since the introduction of these always on, always-connected devices, disallowing use within and adjacent to schools has been the most popular policy. Thus, for a decade or more we have acted as the proverbial ostrich with its head in the sand. Do we really believe that the technology that has brought nearly infinite

Mobile Learning: Perspectives on Practice and Policy
pp. xiii–xvii

learning and knowledge across space and time and into every corner of our lives has no place within the formal organization of learning? We often rue the fact that we perpetuate an educational model birthed in the Enlightenment and only slightly changed by the industrial revolution. Intended to deliver classical knowledge and prepare graduates ready for factory work, this educational model cannot serve the future. And sticking our collective heads in the sand when it comes to mobile learning in and outside of school further perpetuates this obsolete approach to education. Mobile technology is here to stay. We have a wonderfully complex opportunity before us. It is time to get on with it and "make good."

Starry-eyed fantasy is similarly ineffective, so we need to be careful to avoid an overreliance of our historic, hedonistic sci-fi imaginings about the future of mobile learning. We have moved well beyond the first waves of research and practice to see that mobile learning enables something subtler than "learning anywhere," or simply translating existing educational models to a new form of content delivery. In fact, we need to be more jaded than those early cheerleaders because there is no silver bullet for the systemic, interconnected problems mobile learning poses. Mobile communication can be a tool, but is no panacea. The answer to the simple question of whether mobile technology is good or bad for learning is simply, "It's not that simple."

Part of what characterizes the confusion we face when considering mobiles vis-à-vis education is that even the most basic answers are hard to find. For instance, are we clear about what counts as mobile technology? Is a calculator a "mobile device" in the same way as a smartphone? Do screen size, net access, and primary input method differentiate in a meaningful way? What distinctions or similarities account for the depth of difference in software design? Do the same devices we blame for distraction also facilitate ever-deepening focus and access to time-relevant information? If we cannot agree on the boxes, is there a point to studying a thing called "mobile"? Certainly details matter, but it is far from clear which ones, when, and for whom. If we can collectively claim to have learned *one thing* so far about mobile learning, it is how tenuous and tentative our path forward must be to take us anywhere worth going.

Yet, educators and other stakeholders *are* interested in what mobile means for teaching and learning. They are realizing the need to respond to the changes wrought by this technology. Messy and unsure though this ground may be, it is here our authors have sought to find waypoints. Together we have put together a volume that eschews the idealistic gleam or the mostly imaginary worlds so characteristic of earlier work on mobile. Instead, the work in this volume is clothed in realism, zips and buttons of the here-and-now. The authors here engage in the quotidian, hands-on creation of new educational realities and realize that ideals must be tempered by circumstances—including limited time and small, if any, budgets. What we lose in the way of "science fiction," we gain in usefulness to those in more commonly found educational worlds, their questions, and evolving practices.

What is not true of this bewildering landscape conjured by mobile is that it is wholly new. Technology is not new, even if *this* technology is. As a result, we have nearly a century of relevant educational thought to guide us. While not a blueprint for the future, this knowledge helps us see, describe, and work toward education that is more relevant, important, and alive. Perhaps much of our work is looking for ways mobile education can help ensure we maintain the best of 20th century education in our 21st century learning environments. There are so many innovative ideas and practices that were worked out, and worked on, long before mobile technology put phones in so many pockets that have not been outpaced by new inventions. They are as relevant as ever. However, they require us to look from another angle starting with a basic reconsideration of what it means to learn in this new mobile-enhanced world where we have these devices and are constantly connected to networks of others. We have a unique opportunity to rework old paradigms and develop new vigor; thus, our inspiration lies in nurturing the important innovations that have been developing over the past years.

Mobile-based teaching might help us to reconsider inquiry-based learning, where the organizing principle is not the lists of facts and principles to be memorized, but rather nurturing the desire to know and grow. How, for instance, might an inquiry-based model come into its own when every pupil can go off on their own to document and inquire with their mobile devices both connecting them to the inquiry and to the progress of their peers. We find mobile technology requires combining knowledge and experience in new ways. The development of augmented reality, for example, alerts us to the hybridization of place and underscores how these devices are capable of creating new potentials for learning. Another development includes using the onboard technologies (compass, Global Positioning System, accelerometer, camera, advanced processor, etc.) to gather and examine materials in situ. And what about the distribution of basic digital learning materials on inexpensive laptops when, for example, the basic materials of knowledge work, such as a cheap Chromebook, make it possible for a typical school to provide 1:1 resources? While not as flashy as some other technology-based innovations, knowing that students have specific common resources at their disposal opens doors previously closed to many. The basic dynamics of learning occur in an age where knowledge and the ability to create are abundant, and need to be understood and shaped. Participation in knowledge building looks quite different when mobile tools and information access are widely available, not reserved for a few.

Really, a diversity of approaches and perspectives is the only possible strategy when dealing with such wide-ranging and interconnected issues. Like biological evolution, letting a thousand flowers bloom and cross-pollinate looks to be the prudent direction. Just as the centrally planned economies in this world cannot not keep up with the messiness of conditions on the ground, no single group will hit upon a golden mobile learning solution that can be simply scaled. Rather each small success or failure is a contributing insight into what it might mean to succeed or fail in other places and contexts. Sharing, not scaling, is how we reach beyond the here and now. Further, this sharing will not always happen within definitional bounds, such as, among only teachers of seventh-grade math, et cetera. Cross-pollination can come from unexpected corners. An idea needs to find purchase in another plan, and be changed, morphed, and passed on again.

With this book then, we aim to play a constructive role in the evolving ecosystem of knowledge and practices relevant to learning with mobile technology. We have collected the work of educators and researchers in diverse settings who are all concerned, in some way or another, with how we evolve paradigms of teaching and learning in this new age. Readers will see innovative educational approaches that are mediated through the use of new technologies in and outside the classroom. Readers will also hear about efforts to produce policy that is more cognizant of the place these mobile devices can take in learning as well as gain insight into the barriers to spreading the innovative programs to more teachers throughout the world. Of course, professional development's role in broadening the vision and reality of what can be done with mobile devices is crucial. For readers who have a sense that big changes are afoot but have not really seen beyond their own immediate circumstances, this book offers a sense of the bigger picture and the often-unintended consequences and hidden connections involved in planning for learning in the mobile age. Within the chapters that follow it is clear that mobile learning enables not just a better mousetrap, but also a new depth of responsibility to one another.

We have organized the material on mobile learning into three broad sections.

Policy Contexts for Mobile Learning. This section examines issues and questions that reach beyond the level of a single classroom or project, the papers in this section focus on issues of policy, scale, and professional development. It addresses how questions about mobile learning address other big topics, such as whose ideas get time and attention.

Connecting History and Community: Location-Based Mobile Learning. The marriage of mobile technology and game design has resulted in engaging, collaborative learning that can occur both in and outside of schools. The chapters in this section explore place-based games that connect students deeply to their out-of-school communities and histories. This is done by facilitating experiential learning in settings relevant to the content with content relevant to the settings. Whether considered games or just highly interactive educational material, the adult- and student-designed location-based activities showcased here serve as lenses for understanding the educational and technical affordances of mobile technology and how these can be harnessed to scaffold deep, and deeply compelling, learning in and across time and space.

Mobile Learning in PK–12 Educational Settings. Taking the temporal and spatial fixity of the K–12 classroom as a point of departure, this section examines the disruptions and benefits of mobile learning. The authors describe how mobile learning can transform existing classroom interventions, and how they might find new classroom environments within existing school paradigms.

Together, these chapters represent an optimistic take on how we might move forward in today's technologically evolving world. The authors included in this volume see new ways of staking out a future that includes mobile learning. They examine ways to empower learners as designers of their own intellectual destinies, knowing that the path forward means considering pedagogy, policy, and training together. We hope this work lets the reader see some of the promise of mobile learning, as well as the "speed bumps" or even cliffs that lie ahead in what is for so many of us unknown territory. We hope to point toward solutions for learning in a mobile age, not easy ones maybe, but real ones that can grow bigger and better. The authors have done their work and told their stories. We, the editors, have pulled their work together and brought it to you. What these stories become in your lives is their true meaning because, although we have offered perspective, you and those around you will make the future of mobile learning.

PART I

POLICY CONTEXTS FOR MOBILE LEARNING

CHAPTER 1

TENSIONS BETWEEN THE LOCAL IN THE GLOBAL

A Cosmopolitan View of Mobile Learning Initiatives

Judith Dunkerly-Bean and Helen Crompton
Old Dominion University

Char Moffit
California State University, Chico

In this chapter, we utilize the lens of cosmopolitan critical literacy to examine extant NGO policies, sponsored research, and recommendations such as those produced by United Nations Educational, Scientific and Cultural Organization (UNESCO), World Bank, and United States Agency for International Development (USAID) among other nongovernmental organizations (NGOs), to examine tensions between the local and global in mobile learning across three areas: early childhood education, literacy, and teacher professional development. Despite the oft-repeated statistic of more people having access to mobile phones than to indoor plumbing (6 billion and 4.5 billion respectively), the field of mobile learning is still relatively in its infancy and, in some cases, has leapfrogged over the com-

Mobile Learning: Perspectives on Practice and Policy
pp. 3–22

puter revolution (West, 2012). However, along with the unprecedented growth, tensions exist in how best to achieve goals such as eradicating illiteracy by "importing" global technologies while still incorporating local knowledge, context and identities. According to *Education 2030* (UNESCO, 2015), a joint initiative and publication of UNESCO, the World Bank and other organizations, will endeavor to "promote the use of ICT [information and communication technology], particularly mobile learning, for literacy and numeracy programs" (p. 20). Additionally, access to ICT and mobile learning are seen as "essential elements of lifelong learning" (p. 22). Indeed, ICT, in general, and mobile learning, in particular, are seen as integral to literacy and numeracy.

In a comprehensive World Bank report, a similar position is taken in regard to the promise of ICT and mobile learning:

> Information and communication technologies, in particular, have the potential to multiply access to learning opportunities to those who most need them, including hard-to-reach populations, students with disability, and out-of-school youth. Such technologies can also play an important role in improving the quality of teaching—and thus learning—outcomes. (World Bank, 2005, p. 48)

While the goals of global literacy, numeracy, ICT access, and improved teaching practice are laudable and rightfully a global imperative, the question of who decides the content, the access and the approach governing the implementation of mobile learning as an emergent approach remains.

The documents examined here reflect some of the contemporary work by the following NGOs: UNESCO, the World Bank, and USAID. As readers are likely aware, these organizations represent three of the largest and most encompassing NGOs in global operation today, and thus exert considerable influence in initiatives worldwide. While somewhat disparate in their approaches or overarching foci, all are concerned with the staggering number of children and adolescents who are unable to recognize the basic human right of access to fair and equitable education. The latest available data indicates that globally, an estimated 59 million elementary age children and 65 million adolescents, most whom are girls, are currently not in school (UNESCO, 2015). Put in perspective, those numbers would equate to more than a third of the entire U.S. population. Moreover, only 69% of nations were projected to achieve gender parity in education access at the primary level and only 48% at the secondary level (UNESCO, 2015). A renewed focus and commitment to global education was made by UNESCO, the World Bank among other NGOs in 2015 to fulfill the unrealized promise of Education for All in its new iteration,

Education 2030. To familiarize the reader with the three organizations, in the next section we provide a brief description of each.

UNESCO

The United Nations Organization (UN) includes a General Assembly currently representing some 193-member nation states from around the globe. UNESCO is a specialized agency of the UN and works to address issues of global improvements in a wide array of areas related to education. Along with related branches such as widely known UNICEF, UNESCO operates as both a research and implementation entity for global initiatives. Although unable to meet its Education for All goal of compulsory education globally by 2015, UNESCO has put renewed energy, resources and research including the affordances of mobile learning into the goals set by multiple NGOs in Education 2030, which hopes to realize global compulsory education worldwide by 2030.

USAID

Similar in focus, if not in scale to UNESCO, USAID is the primary U.S. organization that, according to their mission statement, works to end extreme global poverty and enable resilient, democratic societies to realize their potential. Current USAID initiatives and investments range from education to global health, conflict resolution to climate change, however given shifts in the focus of U.S. initiatives as of this writing, this may well be impacted. As of this date, however, USAID has not changed its mission to promote global democracy.

The World Bank

The final organization selected for this review is the World Bank. Established in 1944 after the economic devastation of Europe during WWII, the World Bank is one of the world's largest sources of funding and knowledge for development solutions. It is governed by 188 member countries and delivers services out of 120 offices with nearly 15,000 staff located globally. Its current goals include:

- By 2030, ending extreme poverty by decreasing the percentage of people living on less than $1.90 a day to no more than 3%.

- Promote shared prosperity by fostering the income growth of the bottom 40% for every country.

Given the breadth and depth of the work done by these three organizations on a global scale, it seemed appropriate to draw from their extensive body of work in education in general, and mobile learning more recently, for this analysis. Additionally, given the investment of time and international, national and personal resources invested in ICT and mobile learning (Hinostroza, Issacs, & Bourgroum, 2012), it is imperative to build a stronger research base into their use. We begin by situating our review of representative policy documents, research and recommendations within the theoretical framework of cosmopolitan critical literacy.

THEORETICAL FRAMEWORK

As a theoretical perspective, cosmopolitan critical literacy seeks to address the tensions between the global and the local. It asks questions such as, "Who benefits from the construction of race/culture(s) in a particular locale? What identities (global and local) are possible in this locale?" (Dunkerly-Bean, Bean, & Alnajjar, 2014). In the case of our focus here, the cultural and racial contexts may differ widely from place to place, thus demanding attention to the construction and imposition of what it means to be a "literate" individual in a particular locale and context. A cosmopolitan perspective on mobile learning speaks to the possibilities of living and learning at the intersection of the local and the global. Most importantly, a cosmopolitan approach asks us to examine the ways in which mobile learning might contribute to not only betterment of the individual, but also how the learning of the individual might in turn contribute to local/global betterment (Blades & Richardson, 2006; Dejaeghere, 2008; El-Haj & Bonet, 2011). Empirical research is documenting global citizens' literate lives, revealing how a world youth culture is developing, how global migration is impacting youth, how global capitalism is changing their economic and vocational futures, how instant computer-mediated communication with the world is changing the literacy needs and identities of learners of all ages, but especially youth (Hull & Stornaiulo, 2010; Jimenez, Smith, & Teague, 2009; Skerrett, 2012). Research such as this makes clear that the lives of children, adolescents, and adult learners will be lived out at the interface of the local and global (Baildon & Damico, 2011; Hull & Stornaiulo, 2010). In utilizing this framework for this review, we look to examine the global/local tensions in mobile learning, especially as they relate to literacy, early childhood initiatives and teacher professional development and the inclusion of local input in those initiatives.

METHOD OF REVIEW

In the present review, we conducted an analysis of research produced by the largest nongovernmental organizations in global operation today. The rationale for this selection is that these organizations, often in conjunction with large global corporations, are actively involved in initiatives related to developing human and other capital in so-termed "developing countries." In selecting literature for review, we specifically searched for documents, policy briefs, reports, et cetera on the websites and/or linked data bases of these organizations that addressed our area of focus, specifically mobile learning initiatives in literacy, early childhood initiatives, and teacher professional development. This review utilized qualitative content analysis, which is designed to assist the researcher(s) in systematically describing the meaning of large amounts of data.

The overarching goal is to go beyond individual or idiosyncratic understandings of material by addressing the consistency of understanding. For this research, all three authors independently searched for documents related to each of our primary areas of focus. However, we also each read through the shared collection of documents, given the frequent overlap in topic. For example, in some cases, mobile reading initiatives also shared elements of teacher professional development and some early childhood initiatives predictable included elements of reading. Once we had identified the relevant documents we each examined them for evidence of consideration on the part of the organization for local knowledge, voice, and other inputs (i.e., the incorporation of local culture, practices, expertise, and authors.) We turn now to a brief overview of mobile learning in general, before discussing each of our areas of focus: early childhood, mobile reading/literacy, and teacher education and/or professional development in greater depth.

Overview of Mobile Learning

Mobile learning is defined as "learning across multiple contexts, through social and content interactions, using personal electronic devices" (Crompton, 2013a, p. 4.). This definition speaks to the mobility of the device to be used in various locations and across subject areas and that the device is a tool for connecting people and content (text, video, photographs etc.). Vigorous debates have centered on what constitutes a mobile device (Brown, McGregor, & Laurier, 2014; Traxler, 2009). For the purpose of this chapter, Crompton's criteria are used as she postulates that a mobile device should have a prompt on/off button to be available immediately when required and that the device is small enough to be carried and used in the majority of locations, such as sitting on a bus. These criteria include devices such as mobile phones and tablets but exclude laptops.

Smart (Internet enabled) mobile phones have been widely available in many countries for the past decade. However, people living in the Global South typically use basic mobile phones without Internet capabilities. Researchers developing initiatives in the Global South often use a short message service (SMS) to send text messages, as seen in the case of the Self-Employed Women's Association (Nanavaty, 2011) who sent commodity prices to agricultural workers, and also the Jokko project (UNESCO, 2014) who used an SMS network to send various local news items to the participants from birth/deaths, events, and commodity advertising. Indeed, the use of SMS rather than "smart" phone technology is evident in a number of initiatives and speaks to the tensions between global potential and local reality. For example, in 2003, there were several mobile learning initiatives taking place in countries across the globe. These research initiatives typically used a basic mobile phone that had phone and SMS capabilities. Many of these researchers were responding to the call to provide an education to children in developing nations. At the United Nations Millennium Summit in 2000, development goals were written and signed by 189 heads of state from across the world. From those eight goals, one was to provide primary schooling for all children by 2015. The mobile learning work of global agencies has increased greatly since 2011 and those efforts include: (a) drawing together mobile learning researchers and scholars to better understand what we know about mobile learning and how it can be used to extend and enhance learning and providing opportunities to access appropriate learning around the globe, and (b) produce publications to share what is known about mobile learning with policymakers, educational leaders, and other stakeholders (Dunkerly-Bean & Crompton, 2016).

Drawing from those sources, as well as other specific initiative reports, we turn now to the first topic reviewed, that of mobile learning initiatives in early childhood education.

Early Childhood Education

Central to the review of efforts regarding early childhood education through mobile learning are efforts to assess and increase the literacy and numeracy of young children while also providing their caregivers with valuable resources that may otherwise be difficult or even impossible to access or obtain in traditional print form. Two types of projects will be discussed: projects that support early literacy through mobile devices, and projects that feature local practices and culture. As we will see, the two are sometimes mutually exclusive, through efforts to reflect local knowledge appear to be increasing.

One of the primary foci in global efforts involving mobile learning lies in the assessment of early literacy skills. The project eEgra was created by the Education Development Center along with USAID and is utilized in the following countries: Mali, Tajikistan, Kyrgyzstan, and the Philippines. This is a literacy assessment that has been translated into several languages and it can be downloaded for free. Originally a paper-based assessment, it provides teachers with information on a child's reading progress (Education Development Center, 2012). Closely related to this project is the Tangerine/TangerineClass (Beta version) project that is a larger scale project using the assessment in both paper and electronic form to track early literacy learning in 20 developing countries (RTI, 2012). The project Real-time Access and Utilization of Children's Learning Data involves using mobile phones to collect data on a student's literacy learning progress and to guide future instruction in Ghana and Haiti (The Earth Institute, 2014). While it is encouraging to note that theses assessments have been translated into local languages, that is not necessarily the same as representing local knowledge, culture and practice. Aside from the myriad issues with standardized assessment in general, these particular assessments privilege a Global North perspective on what it means to be "literate." Indeed, tasks such as these reify literacy to what can be assessed on a test which includes the largely debunked practice of reading invented or "nonsense" words.

Another widely implemented mobile learning approach for early childhood education involves recorded lessons or literacy activities. These initiatives present some of the more challenging issues with balancing providing resources with the inclusion of local voice. The project MobiLiteracy Uganda is one such example of this. In 2012, Urban Planet Mobile, a company based in Durham, North Carolina was awarded a grant from the All Children Reading competition to pilot a 91-day mobile-phone based program targeting mothers (primarily, but other caregivers of either sex could participate) with simple audio text messages that they could use to support their preliterate children. The audio messages were designed to present all the letters of the Luganda alphabet and to teach the letter sounds and simple words starting with each letter in the context of a simple short story. Children's pre- and postknowledge of the content was measured using an oral assessment based on the Early Grade Reading

Assessment methodology to determine gains. While the local language was used, and arguably increased literacy, it did so with a pedagogical approach that is decidedly Global North in orientation. This became quite evident when it was discovered that parents may have to be away from their children for days or even weeks at a time, although the report did not provide an explanation why. It is also interesting to note that the participant group that was provided with a print version of the curriculum valued it more (i.e., were willing to pay more for it if they were required to) than the mobile version. In reviewing the final report, it is unclear to what extent the authors of the program solicited local education approaches or included local knowledge or practices in developing the initiative and to what degree they may intend to in the future.

Other projects such as the Bridges to the Future Initiative provided primary school students with prerecorded literacy lessons in South Africa and India. Similarly, Broad Class: Listen to Learn provides primary grade children in Pakistan with prerecorded lessons that include a focus on literacy as well as math and health. The primary grade teachers are trained on how to implement the 30-minute daily program in their classrooms (Center for Education Innovations Programs Database, 2013). Another initiative, One Laptop per Child, provided laptops that had games, books, cartoons, and movies to 6-year-old children that had not seen printed text before to see if it would help them with literacy skills in two villages in Ethiopia (Negroponte, 2012). The project, Open Learning Exchange-Teacher Mate provided mobile devices for second- and third-grade students in Rwanda and Ghana to practice English literacy skills for 40 minutes a week. Similarly to Mobiliteracy Uganda, this program utilized curriculum from the United States without any apparent direct solicitation of local input or reflection of local practices or knowledge (Mruz, 2011). An initiative that attempts to blend this "direct-import" model with attention to local contexts is The Radiophone Project. This project creates radio and video versions of the *Galli Galli Sim Sim* program which is the Indian version of the popular American television show, *Sesame Street*. While this program for children ages 2 through 8 in migrant labor communities in India is immediately recognizable to a Global North audience as a *Sesame Street*-type program in design, approach, and format, it provides a core curriculum while providing space for local voice and culture such as music, recognizable characters, dance, and settings (Sesame Workshop, 2012).

However, in all of these initiatives, care must still be taken to actively seek and incorporate local knowledge and agency in pedagogical approaches especially for young learners, rather than assuming that an approach that works in one area of the globe will translate directly and effectively to another, or even be valued in the same way. Likewise, there is a risk of a new colonialism in importing literacy that has at its nexus a Global North perspective. There is a need for more research and a greater effort on the inclusion of local voice and knowledge in the area of mobile learning initiatives in early childhood education.

Mobile Learning for Literacy

While significant gains have been made in combating illiteracy worldwide, it still remains one of the most pressing concerns of the global community. Recent data compiled by the UNESCO Institute for Statistics (UIS) shows that 757 million people, two thirds of whom are women, still lack basic skills required to read and write a simple sentence (UNESCO, 2013). The UIS reports that the increases in global literacy rates over the past several years are largely due to increases or revised reporting in three countries: India, the Democratic Republic of Congo, and the United Republic of Tanzania. Despite gains or more accurate reporting in the above-mentioned nations, other countries such as Pakistan, Senegal, and Uganda are reporting increases in illiteracy rates, again with the majority of the illiterate being women and girls. Thus, despite an overall increase in literacy from 2000–2015, only 39 countries are considered to be likely to meet the Education for All goal of a 50% decrease in adult illiteracy rates (UNESCO, 2015).

With initiatives such as Education for All and the current Education 2030, it is clear that NGOs such as UNESCO, the World Bank, and USAID are searching for expedient and effective methods for combating illiteracy, in general, and providing equitable access to books and literacy for women and girls in particular. In this section, we will examine the major research, reports and resulting policy produced by these three entities for the inclusion of local voice and identity in the initiatives most widely and currently implemented involving mobile reading.

If mobile learning in global contexts generally is a relatively new phenomenon, then mobile reading as an approach to addressing global illiteracy is still in its infancy. Research into its implementation and efficacy are not robust, yet early reports from involved NGOs are promising. To capitalize on the existence of even basic mobile phone ownership and usage, mobile phone corporations partnered with UNESCO and the nonprofit organization World Reader to utilize a mobile reading app to provide access to text to those for whom print based books are impractical, expensive or simply inaccessible. Worldreader Mobile is an application that allows people to access books and stories from a wide variety of mobile phones, including inexpensive feature phones. To determine the effectiveness of mobile reading in developing areas, UNESCO conducted a large survey of mobile reader users to determine who was reading, what they were reading, and how the users of mobile reading felt about the experience of reading on mobile device (West & Chew, 2014).

That research found that, on average, Worldreader Mobile had 334,000 active users per month. To utilize it, users download the free application, which is stored in the memory of the phone. It is important to note that, while Worldreader Mobile provides access to more than 6,000 digital titles, mobile data connection and the resulting expense is the responsibility of the user. Although these costs are relatively low by Global North standards they can be prohibitive for those in so-termed developing countries and economies. Likewise, despite the fact that a majority of the titles available through the app are free, most of the books are not downloadable, nor can they be read offline further problematizing access issues. However, results also indicated that the application provided opportunity for women to read more widely than they might have otherwise due to the relative privacy of reading on a phone rather than a book. Surprisingly, they also found that more men than women utilized Worldreader Mobile to read to their children (traditionally and culturally a maternal practice). However, this result, while encouraging, is somewhat mitigated by the fact that more men than women owned phones to begin with. Clearly, this is an area requiring further study to determine if, and to what extent, mobile reading is influencing cultural practice.

Another initiative, iRead Ghana, provided e-readers to public school students and teachers, categorized as a Global Development Alliance program between USAID and the aforementioned Worldreader project as a nonprofit organization. This study gave purposively sampled (n = 481) Ghana public-school students access to books through e-readers. While pre- and postliteracy attainment was not an objective of this study, the findings for this initiative indicate that the majority of students and teachers participating in iREAD Ghana had positive experiences with the e-reader and expressed the belief that there is a place for digital media and e-reader technology in the Ghanaian school system (USAID, 2012).

However, from the perspective of soliciting or including local voice, the report had less than glowing comments for the local dialect and oral traditions, which they deemed sociocultural limitations:

> *English language limitations*: While English is the official language of Ghana, it is the second language for most if not all project-affected students and teachers. English language limitations were especially significant for primary students who use English for the first time in primary class 4.
>
> *Potentially unsupportive home environments*: As many project-affected students are from agricultural families, their home environments may not appreciate reading as an activity. For example, students may spend more time at home performing family chores and agricultural tasks rather than

> engaging with their e-readers. Additionally, students whose parents have limited English and literacy skills may be unable to actively incorporate the device into home life. (p. 22)

It is salient to our argument here that, in this report, agricultural tasks and knowledge are not seen as "literate" and parents with limited English are framed in the language of deficit, rather than acknowledging the myriad of literacies involved in agricultural work or the time constraints for recreational activities inherent in this context. Moreover, the report takes to task the emerging efforts of Worldreader to provide books in the local language. Indeed, the final report disparages that the books may not be in what the report authors considered "standard" English:

> Findings within this report demonstrate that many students had an affinity toward African reading material because it was representative of their culture. A concern, however, is that many of the locally authored texts had grammatical, syntactical, and typographical errors that do not present a positive model for young readers. (p. 24)

To surmise that there is something lacking in wanting to read books in their own language and that reflect local culture is indicative of the perils of neocolonialism even within global literacy efforts.

Another similar initiative, Project LEAP (Jaffe, Lowe, & Tam, 2014) tested the use, function and adoption of e-readers in participating libraries in an effort to understand and investigate the potential and feasibility of library e-reader programs across Kenya and sub-Saharan Africa. One of the stated goals of this initiative was to increase reading through libraries, and therefore improve literacy skills and ameliorate the effects of illiteracy on poverty, health, gender equality, and social and economic stability. Worldreader contributed 200 e-readers (each preloaded with 200 digital book titles) to the selected pilot libraries, for a total of 40,000 books.

Findings from this pilot program included concerns about the diversity and size of library collections and the need for more technology programs as priority concerns prior to the e-reader program initiative. In addition, it was found that young people (determined in the study to be under the age of 25) were the primary users of the e-readers, and points to an important need to engage both youth and older citizens. However, perhaps most importantly and related to issues of access and availability, both library patrons and staff stressed the importance and value of social interactions and the ability to discuss what was being read along with obtaining technical assistance when required. That is, the culturally relevant practice and emphasis on oral tradition needed to be included in the infusion of technology and mobile reading. While Worldreader has partnered with more than 100 African authors to digitize their work, the demand is still outpacing the supply and speaks to concerns around who has the power to determine what is available to readers in the developing world. Even with these efforts and perhaps due to the process of digitizing texts, the proportion of local texts to other titles remains relatively small. Lastly, in these initiatives there was somewhat of a cultural "disconnect" between views on the value and place for printed text, valued highly in the Global North, and in societies where social interaction and oral traditions are highly valued. This is in keeping with Totemeyer's (1994) observation:

> Books and libraries are often seen as redundant in societies that are mainly based on oral traditions and practices. In such societies, people stop reading once formal education is completed as they derive more pleasure from the oral and performing arts—talking, singing, dancing, socializing—than from the rather private and individual reading of a book. (p. 54)

Although attractive from the perspective of the Global North, the addition of technology in the form of e-readers may not take into consideration local values and practices. We turn now to perhaps one of the most crucial aspects of mobile learning initiatives—teacher education and professional development.

Teacher Education and Professional Development

During the September 2000 Millennium Summit, world leaders gathered to finalize and adopt the UN Millennium Declaration. This declaration provided evidence of the commitment these nations were making to work together to meet set goals that will support positive global change. These goals were named the Millennium Development Goals and a target was set for reaching these goals in 2015. From these eight goals, the second goal was to provide universal primary education, to ensure that, by 2015, children everywhere, boys and girls alike, will be able to complete a full course of primary schooling. From the Millennium Development Goals Report published in 2015 (Millennium Project, 2015), the reports show that the goals were mostly met. The primary school net enrollment rate in the developing regions moved from 83% in 2000 to 91% in 2015. The number of out-of-school children of primary school age worldwide fell by almost 50% to an estimated 57 million in 2015, from 100 million in 2000. In places where primary education was greatly lacking, such as in sub-Saharan Africa, the number of children enrolled in primary schools more than doubled from 62 to 149 million.

The work of these partner countries has made tremendous strides in achieving universal primary education. However, accomplishing this goal has led to new problems that can also lead to the undoing of all the positive work that has been achieved. The problem that has appeared is that there are not enough teachers for these new students. As reported by UNESCO, there are now students in the classrooms, but there is also a great shortage of teachers to work with these students (Park, 2016).

In many countries, such as South Africa, there is not the infrastructure in place to provide those teachers and with the tremendous pressure placed on the existing teachers, many of those are also leaving the profession (Van Broekhuizen, 2015). Gaining teacher training statistics in South Africa is difficult, but using aggregated data from the Higher Education Management Information System, Van Broekhuizen conducted a study to learn more about the lack of teachers in this country. From this work, he found that although there was a rise in those gaining teacher training, there were not enough teachers to meet the demand.

In addition, Van Broekhuizen (2015) found a that the teachers lacked many of the necessary skills they are required to use and are often required to teach what they are not trained to teach. As Broekhuizen reports, "Unfortunately, while there is little systematic information on teacher utilization in South African schools, qualitative evidence suggests that many teachers are employed to teach subjects for which they are not trained" (p. 84). With the lack of appropriate training and large increase in the number of students, Higher Education Management Information System data show that between 20,000 and 30,000 teachers are required per year to teach and replace those leaving the profession.

One strategy used is to bring in unqualified teachers, and there are a reported 10,000 such unqualified teachers working in South Africa (Hawker, 2013). As the education department spokesman declared,

> The country had an option. Do we bring learners to the school to sit in an empty classroom where there is no teacher, or [can] you get somebody who can teach them and give support to that particular teacher? And we chose the latter option. Getting someone who can be in the classroom and at least teach learners and we then give support. (Reported by Hawker, 2013)

In 2015, world leaders again met to determine goals for the way forward in the plight to mobilize efforts to end all forms of poverty, fight inequalities, and tackle climate change, while ensuring that no one is left behind. The Sustainable Development Goals were developed from this meeting and of the 17 goals, Number 4 was to ensure inclusive and quality education for all to promote lifelong learning. One of the main targets within this SDG is to substantially increase the support of qualified teachers through teacher training in developing countries by 2030.

Providing teacher training and professional development is not an easy task to organize. The lack of material resources and local experts cause difficulty in obtaining this training to supply the teachers needed and to better equip the teachers already working in these countries. Many researchers have used mobile devices as a means to provide that training and professional development. Mobile devices, such as the basic phone, are lightweight and do not require bulky transit. Experts can be used to provide the training using audio or SMS to carry the training needed to the teachers. Even in areas without electricity, the devices can be powered in the cities and last for days with the stored power supply in the batteries. Thus, even basic mobile phones can be utilized to provide access to professional development and education opportunities in a range of settings. For example, the Text2Teach initiative used basic mobile phones for teachers to access mathematics, English, and science videos. The teachers could watch these short videos to better understand certain concepts and or could play them on television sets or portable projectors for the children in class. The Text2Teach initiative began in 2003 in the Philippines and has expanded to Colombia, India, Chile, and Nigeria. Funded by Globe Telecom, Nokia, the Pearson Foundation, and the Ayala Foundation, this initiative has trained more than 3,555 teachers and school administrators worldwide (Cristanto, 2016).

In another initiative, Learning Communities Enabled by Mobile Technology: A Case Study of School-Based, In-Service Secondary Teacher Training in Rural Bangladesh, researchers used mobile devices to provide training to 20 Bangla and math teachers in 10 schools of the Barisal region in southern Bangladesh (Pouezvara & Khan, 2007). The results of the study show that the teachers appreciated the ongoing communication between the trainer and the teacher that the mobile devices provided. The teachers also liked that they did not have to leave their families to experience the training, rather the mobile devices enabled them to have the training in their own homes.

More recently, Mxit, an SMS messaging system, has been used to provide training to teachers in South Africa (O'Hagan, 2011). Mxit is a popular social networking platform with 10 million subscribers. In November 2012, Mxit reported 5 million subscribers to its educational content and 600,000 to its eight exam preparation applications Dr. Maths is an online tutoring system on Mxit that enables 32,000 learners from across the country to ask questions, chat, and gain real-time support from the qualified math tutors accessible using Mxit (eLearning Africa, 2012).

Taken together, these initiatives provide a thought-provoking example of the nexus between the local and the global. On the one hand, all of these initiatives are designed to address local issues related to teacher education and professional development. Each are selected for implementation by local schools, districts or departments/ministries of education. In that sense—local input is profound. Indeed, the success of these initiatives is highly dependent of local "buy-in" and implementation.

However, just as in the case of early childhood and literacy efforts, there are tensions between global organizations and corporations imposing a vision or model for teacher education and development that may not take into full consideration local expertise or knowledge. Rather, it seems in this area as well that there exists a privileging of a Global North perspective on what it means to be a teacher. In some cases, this means being able to deliver, but not contribute to the development of curriculum. One notable exception to this, and perhaps a model for other initiatives, is Text2Teach. As described above, the initiative is designed to assist teachers in developing mastery in challenging content areas. In this instance, however, the notable difference is that Text2Teach was developed in the Philippines as their version of the global BridgeIT program. Thus, a global initiative was interpreted for a local context, which capitalizes on local knowledge and expertise.

CONCLUSION

In this chapter, we have attempted to illustrate the ways in which mobile learning projects have incorporated local contributions in global NGO initiatives. Drawing from the lens of cosmopolitan critical literacy we examined the tensions of the interplay between the local and the global. While these initiatives are rooted in the desire to provide greater access to information, learning, literacy and professional development, there remains a risk of privileging a Global North perspective over the local knowledge, customs, and context. Additionally, given that language and literacy are primary markers of a particular culture, there is often an uneasy tension between globally predominant languages such as English, and local languages and/or dialects. Indeed, education and enculturation are often synonymous with each other. Thus, there is a fine line between offering new technologies to be incorporated into existing cultures and social systems and imposing new technologies in a manner that may subvert or replace such cultures and systems. As Traxler (2013) succinctly argues:

> These technologies project the pedagogies, strictly speaking perhaps the epistemologies, of outsiders into communities that of course already have their own learning. There is a risk that mobile technologies delivering learning in this way represent either a Trojan horse or a cargo cult that threatens or undermines a fragile educational ecosystem. The issue is not one of emerging markets or developing regions per se but of fragile cultures (or subcultures or even counter cultures) and their capacity to negotiate an optimal balance between the preservation of language, heritage and culture

> on the one hand and engagement with the wider world and the global knowledge economy, on the other. (p. 49)

It is in that delicate balance of local preservation and global engagement that may present the biggest challenge as we move forward with expanding the reach, breadth, and pace of mobile learning initiatives. Fortunately, some models of incorporating the local and the global are already in operation. For example, the project, 1001 Stories found in Uganda and India through the organization Seeds of Empowerment, involves volunteers who teach children about storytelling in order to tell their own stories. The best stories are then sold through the Amazon and Apple apps to be able to continue to fund the projects (Seeds of Empowerment, 2012). Although this approach may seem to be neoliberal or capitalist, it does allow for local engagement to continue beyond the first cycle of intervention.

Similarly, SenMobil—Kindergarten Mobile Learning provides kindergarten teachers in Senegal with locally made mobile apps on phones that support literacy and math learning to aid teachers in their content knowledge and pedagogy, as local kindergarten classes tend to be quite large (SenMobile, 2012). Likewise, The Mobile and Immersive Learning for Literacy in Emerging Economies was created to bring literacy learning games to young children in India that were developed through Carnegie Mellon University with the input and assistance of local English language learners teachers who were consulted to determine local needs (Human Development Lab Carnegie Mellon University, n.d.). Initiatives such as these may provide a model for addressing pressing global educational issues while still equally valuing local voice and agency.

Intrinsic to efforts like those outlined above is the recognition of the multifaceted interconnections between global flows, particularly of information and ideas, when digital media "travel" from one locale to another, far removed not only in place but in resources and power as well. From a cosmopolitan critical literacy perspective, key issues related to mobile learning and digital media include who decides the content to be shared, how to share it and to what extent (if any) is local knowledge and context solicited and represented in the content? In other words, who is represented and who is (even inadvertently) silenced?

In the case of "bring your own technology," or individuals selecting which apps to use, how, and why, the question may not be as important although still worth considering. However, in the case of initiatives where mobile readers, recorded lessons or curriculum written in the Global North are mobilized for uses in far-removed locales, there does seem to be a pronounced risk of imposing a Global North perspective, rather than taking into consideration local perception and input. Similarly, questions of perception regarding the "developer" and the "developing" may inadvertently reascribe age-old power relationships between North and South. It is in that delicate balance of honoring local culture and practice with global initiatives originating in the North that may present the biggest challenge for mobile learning as we move forward. In doing so, it is valuable to remain cognizant of the situated and contextualized use of mobile devices in ways that may further illuminate the affordance and challenges of balancing the global with the local.

REFERENCES

Baildon, M., & Damico, J. S. (2011). *Social studies as new literacies in a global society: Relational cosmopolitanism in the classroom*. New York, NY: Routledge.

Blades, D. W., & Richardson, G. H. (2006). Restarting the interrupted discourse of the public good: Global citizenship education as moral imperative. In G. H. Richardson & D. W. Blades (Eds.), *Troubling the canon of citizenship education* (pp. 115–123). New York, NY: Peter Lang.

Brown, B., McGregor, M., & Laurier, E. (2014). iPhone in vivo: Video analysis of mobile device use. *Proceedings of CHI 13* (pp. 1031–1040). Paris, France: ACM Press.

Cristanto, Y. (2016). An MNO's perspective: A decade of implementing mobile learning in primary schools in the Philippines [Presentation]. Retrieved from http://www.unesco.org/education/MLW2014/R3WAMB2-YolandaCrisanto-Globe.pdf

Crompton, H. (2013a). A historical overview of mobile learning: Toward learner-centered education. In Z. L. Berge & L. Y. Muilenburg (Eds.), *Handbook of mobile learning* (pp. 3–14). Florence, KY: Routledge.

Crompton, H. (2013b). Mobile learning: New approach, new theory. In Z. L. Berge & L. Y. Muilenburg (Eds.), *Handbook of mobile learning* (pp. 47–57). Florence, KY: Routledge.

Dejaeghere, J. (2008). Citizenship as privilege and power: Australian educators' lived experiences as citizens. *Comparative Education Review, 52*(3), 357–380.

Dunkerly-Bean, J. M., Bean, T. W., & Alnajjar, K. (2014). Seeking asylum: Adolescents explore the crossroads of human rights education and cosmopolitan critical literacy. *Journal of Adolescent & Adult Literacy, 58*(3), 230–241.

Dunkerly-Bean, J., & Crompton, H. (2016). The role of mobile learning in promoting global literacy and human rights for women and girls. In B. Guzzetti & M. Lesley (Eds.), *Handbook of research on the societal impact of digital media* (pp. 582–609). Hershey, PA: IGI Global.

The Earth Institute. (2014, January 15). *Real-time access and utilization of children's learning data: Case study from the Millennium Village Project* [video file]. Retrieved from http://www.meducationalliance.org/content/featured-video-real-time-access-and-utilization-childrens-learning-data-earth-institute-colu

Education Development Center. (2012). *What is eEGRA?* Retrieved from http://eegra.edc.org/en/about/what-is-eegra/

eLearning Africa. (2012). Mixing it with Dr. Math: Mobile tutoring on demand. Retrieved from http://www.elearning-africa.com/eLA_Newsportal/mixing-it-with-dr-math-mobile-tutoring-on-demand

El-Haj, T. R. A., & Bonet, S. W. (2011). Education, citizenship, and the politics of belonging: Youth from Muslim transnational communities and the "War on Terror." *Review of Research in Education, 35*, 29–59.

Hawker, D. (2013). Thousands of unqualified teachers are teaching SA children. eNews Channel Africa. Retrieved from https://www.enca.com/south-africa/thousands-unqualified-teachers-sa-schools

Hinostroza, J. E., Isaacs, S., & Bougroum, M. (2012). *Information and communications technologies for improving students' learning opportunities and outcomes in developing countries* [Brookings GCL Draft Working Report]. Washington, DC: Brookings Institute.

Hull, G. A., & Stornaiulo, A. (2010). Literate arts in a global world: Reframing social networking as cosmopolitan practice. *Journal of Adolescent & Adult Literacy, 54*(2), 85–97.

Human Development Lab Carnegie Mellon University. (n.d.) Mobile and immersive learning for literacy in emerging economies (MILLEE). Retrieved from http://www.cs.cmu.edu/~mattkam/lab/millee.html

Jaffe, S., Lowe, Z., & Tam, T. (2014). Project LEAP: Libraries, e-reading, activities, partnership baseline report. Retrieved August 18, 2014 from http://cdn.worldreader.org/wp-content/uploads/2014/08/Project-LEAP-Baseline-Report_2.pdf

Jimenez, R., Smith, P., & Teague, B. (2009). Transnational and community literacies for teachers. *Journal of Adolescent & Adult Literacy, 53*(1), 16–28.

Millennium Project. (2000). Millennium goals. Retrieved November 15, 2016, from http://www.unmillenniumproject.org/goals/

Mruz, K. (2011). *2011 Rwanda TeacherMate Report.* Cambridge, MA: Open Learning Exchange. Retrieved December 9, 2013, from http://www.innovationsforlearning.org/resources/content/8/9/documents/Rwanda_TeacherMate_Reports.pdf

Nanavaty, R. (2011). The value of information: Mobile technology helps microentrepreneurs. *Digital Development Debates*. Retrieved March 20, 2016, from http://www.digital-development-debates.org/issue-06-innovation--science-and-technology--the-value-of-information-mobile-technology-helps-micro-entrepreneurs.html

Negroponte, N. (2012). EmTeach preview: Another way to think about learning. *MIT Technology Review.* Retrieved from http://www.technologyreview.com/view/429206/emtech-preview-another-way-to-think-about-learning/

O'Hagan, T. (2011). Mobility in education: Can mobile devices support teaching and learning in South Africa? Retrieved from hsf.org.za/resource-centre/focus/focus-68/(11)%20T.%20OHagan.pdf/download

Park, J., Tan, M. M., & Teng, J. X. (2016, March). *Policies to improve children's safe, effective and responsible use of ICT: A review of the Asia Pacific*. Paper presented at UNESCO mobile learning week, Paris, France.

Pouezvara, S. L. & Khan, R. (2007). Innovative information and communication technology in education and its potential for reducing poverty in the Asia and

Pacific Region learning communities enabled by mobile technology: A case study of school-based, in-service secondary teacher training in rural Bangladesh. Retrieved from https://www.researchgate.net/publication/237366658_Innovative_Information_and_Communication_Technology_in_Education_and_Its_Potential_for_Reducing_Poverty_in_the_Asia_and_Pacific_Region_Learning_Communities_Enabled_by_Mobile_Technology_A_Case_Study_of

RTI International. (2012, January). Kenya field trial report: Summary. Tangerine—Electronic data collection tool for early reading and math assessments. Retrieved from http://www.rti.org/files/Tangerine_report_0112.pdf

Seeds of Empowerment. (2012). 1001 stories. Retrieved from http://seedsofempowerment.org/projects/1001-stories-2/

SenMobile. (2012, May 24). Introducing mobile learning in kindergarten in Senegal [Video File]. Retrieved from http://www.youtube.com/watch?feature=player_embedded&v=BHw9SOaaOGCg

Skerrett, A. (2012). Language and literacies in translocation: Experiences and perspectives of a transnational youth. *Journal of Literacy Research, 44*(4), 364–395.

Sesame Workshop. (2012, August 8). Helping life-saving lessons reach marginalized Indian communities. Retrieved from http://www.sesameworkshop.org/our-blog/2012/0808/helping-life-saving-lessons-reach-marginalized-indian-communities/

Totemyer, G. (1994). Challenges for democracy: Decentralization and empowerment in Africa. *Regional Development Dialogue, 15*, 49–61.

Traxler, J. (2009). Learning in a mobile age. *International Journal of Mobile and Blended Learning, 1*(1), 1–12.

Traxler, J. (2013). mLearning: Solutions for international development—Rethinking the thinking. *Digital Culture & Education, 5*(2), 74–85.

Urban Planet Mobile. (2012, September 19). USAID awards urban planet mobile with a grant to develop mobile educational program in Uganda. Retrieved from http://www.urbanplanetmobile.com/news/usaid-awards-urban-planet-mobile-with-a-grant-to-develop-mobile-educational-program-in-uganda

UNESCO. (2013). Adult and youth literacy [Fact sheet]. Montreal, Quebec, Canada, UNESCO Institute for Statistics.

UNESCO. (2014). Jokko Initiative. Retrieved from http://www.unesco.org/uil/litbase/?menu=4&programme=181

UNESCO. (2015) Education 2030. Incheon Declaration. Retrieved from http://www.unesco.org/new/fileadmin/MULTIMEDIA/HQ/ED/ED_new/pdf/FFA-ENG-27Oct15.pdf

USAID. (2012). iREAD Ghana study: Final evaluation report [Report]. Retrieved from http://www.worldreader.org/uploads/Worldreader%20ILC%20USAID%20iREAD%20Final%20Report%20Jan-2012.pdf

Van Broekhuizen, H. (2015). Teacher supply in South Africa: A focus on initial teacher education graduate production. Stellenbosch Economic Working Papers: WP07/15. from www.ekon.sun.ac.za/wpapers/2015/wp072015/wp-07-2015.pdf

West, M., & Chew, H. E. (2014). Reading in the mobile era: A study of mobile reading in developing countries [Report]. Retrieved from http://www.unesco.org/new/en/unesco/themes/icts/m4ed/mobile-reading/reading-in-the-mobile-era/

World Bank. (2005). Education Sector Strategy Update. Retrieved from http://siteresources.worldbank.org/EDUCATION/Resources/ESSU/Education_Sector_Strategy_Update.pdf

CHAPTER 2

THE ROLE OF LEADERSHIP AND PROFESSIONAL DEVELOPMENT TO OVERCOMING BARRIERS TO MOBILE LEARNING IN FORMAL SCHOOLING

Liz Kolb
University of Michigan

Utilizing mobile devices for school learning has several potential benefits. Examples of these benefits include: opening up and enriching educational experiences in diverse settings, building communities of learners, situating learning in authentic contexts, enabling seamless learning, providing more equitable opportunities for students to learn, facilitating personalized learning, providing immediate feedback and assessment, and enabling anytime, anywhere learning (Murray & Olcese, 2011; Policy Guidelines for Mobile Learning, 2013; Shuler, 2009). Understanding the possible affordances of mobile learning, countries such as the Australia, United States, Japan, and some countries of the European Union, have been promoting mobile learning as part of their national education

Mobile Learning: Perspectives on Practice and Policy
pp. 23–43

policies (Hylén, 2012; Khan, Al-Shihi, Al-Khanjari, & Sarrab, 2015; U.S. Department of Education, Office of Educational Technology, 2016). In both developing and developed nations, teaching with mobile devices requires an extraordinary amount of preparation, planning, and policy support (Hylén, 2012; Muir, 2013). Even in the most industrialized countries there are significant challenges to successfully implementing mobile learning initiatives to fulfill the benefits of these programs. Some of these challenges include pedagogical, technical, policy, cost, infrastructure, leadership, digital citizenship and cyber safety, assessment and research (Baker, Dede, & Evans, 2014; Khaddage et al., 2015; Policy Guidelines for Mobile Learning, 2013). This chapter considers aspects of policy that are often overlooked when initiating mobile learning programs in formal school settings. While this chapter concentrates on North American and European policies toward mobile learning, the recommendations can be transferred to school systems across the globe.

Historically, mobile learning has been defined as broadly as interactions among a wide array of mobile devices, the social environment, and the learners (e.g., Crompton, 2013; Koole, 2009; Policy Guidelines for Mobile Learning, 2013), and as narrowly as the sole use of a particular mobile device for learning, such as a smartphone or tablet (Quinn, 2000; Traxler, 2005). For this chapter, mobile learning will use the broader definition from United Nations Educational, Scientific and Cultural Organization (Policy Guidelines for Mobile Learning, 2013), namely,

> Mobile learning involves the use of mobile technology, either alone or in combination with other information and communication technology (ICT), to enable learning anytime and anywhere. Learning can unfold in a variety of ways: people can use mobile devices to access educational resources, connect with others, or create content, both inside and outside classrooms. Mobile learning also encompasses efforts to support broad educational goals such as the effective administration of school systems and improved communication between schools and families. (p. 6)

While this chapter will use this broad definition of what constitutes mobile learning and the devices utilized, the chapter takes a targeted focus on mobile learning policies concentrating on formal school settings.

A FOCUS ON NORTH AMERICAN AND EUROPEAN SCHOOL SYSTEMS

Government agencies in Canada and the United States have encouraged local school systems to include mobile learning as an integral part of their long-term planning (Learning and Technology Policy Framework, 2013;

Ministry of Education Transformation Plan, 2013; U.S. Department of Education, Office of Educational Technology, 2016). In Europe, countries such as the Netherlands, Denmark, and the United Kingdom are endorsing mobile learning as part of their policy in formal education (Hylén, 2012). Despite national support for schools to integrate mobile learning, there is not yet agreement on how mobile learning should be and can be sustained in more formal school settings (Ng & Nicholas, 2012). Furthermore, Ng and Nicholas (2012) argue that without policy to ensure sustainability for mobile learning programs in education institutions, investments will be wasted and future innovations may be threatened.

School systems struggle with developing mobile device policies that are consistent across classrooms and include considerations for various learner needs, pedagogical approaches, financial costs, infrastructure upgrades, equity concerns, social attitudes and teacher needs. As of 2014, around 30% of primary and secondary schools in the United States have a one-to-one policy where the school provides mobile devices to each student (The Digital Playbook, 2014). In addition, 51% of school districts either allow or plan to allow students to bring their own devices to school (bring your own device or BYOD) to use for learning (The Digital Playbook, 2014). Across Europe, an average of 75% of schools have BYOD policies (Blamire & Colin, 2015). Yet, only 38% of these schools are actually providing services to support the inclusive BYOD policies (Blamire & Colin, 2015). This large number of schools with inclusive BYOD policies that lack services to support BYOD demonstrates a problematic situation. When schools first develop a mobile learning program, the focus tends to be on the type of mobile devices and on developing rules for appropriate, secure and safe use of the physical devices. While legal issues and devices damage are important, they are not the only concerns that schools need to be addressing. Studies have shown that in mobile programs, the type of devices used was less a determining factor in sustained academic successes than having a systematic buy in and vision taken by successful schools (Keane & Keane, 2017). Often forgotten are large-scale policy issues should also be implemented in order to have a successful and sustainable mobile device program (Valiente, 2010).

School systems have cited a number of challenges to a successful mobile program.[1] Despite the recent push from federal and local governments, some schools have abandoned their one-to-one mobile programs because of lack of technical support, the financial strain of maintaining the program, problems with student distraction, or lack of effective positive learning outcomes (Barshay, 2014; Fried, 2008; Hu, 2007; Lemagie, 2010; Warschauer, 2006). Education organizations must take a holistic approach to mobile learning in order for it to be successful, which includes not only infrastructure to support the program but also to plan for how digital tools intersect with learning as well as the society around the learning institution (Cervantes, Warschauer, Nardi, & Sambasivan, 2011; Policy Guidelines for Mobile Learning, 2013). This chapter will address and provide recommendations for the marginalized, yet significant areas of policy that formal educational institutions need to consider for long-term successful mobile device implementation (Baker et al., 2014; Hylén, 2012). This paper argues that the two most important overlooked areas of mobile program planning are leadership and teacher professional development: both playing a substantial role in determining device policy, infrastructure, community partnership, equity needs, digital citizenship and cybersafety, student data privacy, infrastructure, and program assessment.

COMMON APPROACHES MOBILE LEARNING DEFINED

This chapter assumes there are four common approaches to mobile learning in these settings in schooling: one-to-one, one-to-many, BYOD, and hybrid. The benefits and drawbacks of these different mobile approaches are described in Table 2.1. Below are the definitions of these four mobile learning programs.

- One-to-One: The school system provides one mobile device for every student. The device may be kept at school or some schools allow students to take devices home. The device is often purchased or leased through school funds. Occasionally schools will require families to purchase or lease the device. This approach is more common in schools of lower socioeconomic status (The Future of Mobile Learning, 2013; McLean, 2016).
- One-to-Many: Schools provide devices but not enough for every student to be assigned to their own device for the year. Rather, there are often carts or cases of mobile devices that instructors can check out and use during school for instruction with the students (McLean, 2016).
- BYOD or BYOT (bring your own technology): Schools allow students to bring in and use their own mobile devices from home. These devices vary widely from iPods to smartphones to iPads to handheld video game systems to laptops. This approach tends to be more common in schools in higher socioeconomic areas. While BYOD/BYOT tends to be less costly, it does put additional pressure on the school's bandwidth and teachers to support a wide variety of devices coming into the classrooms (Fallon, 2015; The Future of Mobile Learning, 2013).

Table 2.1. The Benefits and Drawbacks of Different Mobile Learning Approaches

Type of Approach	*Affordances of Approach*	*Limitations of Approach*
One-to-one	• Each student has same device with identical software • Easier to troubleshoot and maintain devices • Easier for teachers to learn one device • Students often have access to the device 24/7 • Opportunities for personalized learning with device • Potentially opportunities to connect everyday experiences with school learning • Equity: Parents have access to a digital tool that they may not have access to (allows potential for knowledge and educational growth in whole family)	• Costly for schools to sustain over long periods of time • Training needed for families to help support the technology use at home • Vast infrastructure needed both inside and outside of school to maintain devices (tech support, WiFi and bandwidth, etc.) • Pressure or expectation that all teachers will be using the technology • Equity: It may be difficult for some students to complete work outside of school that may require Internet access
One-to-many	• Less costly than a one-to-one program • Each student has same device with identical software • Easier to troubleshoot devices • Easier for teachers to learn the device	• Students do not have access to the device 24/7 • Can be difficult to keep track of student work, especially if they do not always use the same device • Equity: There are concerns with access to devices to connect to the Internet outside of school • It may be difficult for some students to complete work outside of school that may require a device and/or Internet access

(Table continues on the next page.)

Table 2.1. (Continued)

Type of Approach	*Affordances of Approach*	*Limitations of Approach*
BYOD	• Less costly than one-to-one or one-to-many. • Students have access to the device 24/7. • Potentially opportunities to connect everyday experiences with school learning. • Students use a device that is most comfortable to them.	• Many different types of devices and applications being used in the classroom • School may have to supplement devices • Strong bandwidth needed to support all the devices on the network • Teachers may not be comfortable with the variety of devices and lack of control over what is on the devices and how they work • Students can often access unfiltered Internet networks through their own mobile devices • Student data, security, and privacy are harder to monitor and protect • Equity: There are concerns with access to devices to connect to the Internet outside of school. It may be difficult for some students to complete work outside of school that may require a device and/or Internet access
Combination (BYOD and one-to-many)	• Less costly than one-to-one or one-to-many • Students are able to use the device of their choice • All students will have access to a device in school • Teachers could use the school devices for some activities for continuity of applications if they prefer not to use BYOD • Equity: All students should have access to some type of mobile device in school to complete their assignments	• Many different types of devices and applications in the classroom • Teachers may not be comfortable with the variety of devices and lack of control over what is on the devices and how they work • Students can often access unfiltered Internet networks through their own mobile devices • Student data, security, and privacy are harder to monitor and protect • Equity: There are concerns with access to devices to connect to the Internet outside of school. It may be difficult for some students to complete work outside of school that may require a device and/or Internet access

Sources: Based on data from Digital Equity Action Toolkit (2016); The Future of Mobile Learning (2013); Hylén (2012); Policy Guidelines for Mobile Learning (2013); Lemagie (2010); and Vascellaro (2006).

- Hybrid (BYOD and one-to-many): Schools participate in both providing devices for students at school and allowing them to bring in their own devices. Most often this approach is used by districts that cannot afford a full one-to-one program and do not have as many students with their own robust devices.

VISIONARY LEADERSHIP

While a school clearly needs a strong and sustained digital network to support the technology (infrastructure will be discussed later in the chapter), no mobile program is successful based on its infrastructure alone (Keane & Keane, 2017; Smith, 2008). Nontechnical issues are just as important as technical issues to a successful implementation of mobile programs (Holcomb 2009; Tedre, Hansson, Mozelius, & Lind, 2011). Fundamental to success is the school leadership creating a collective vision for mobile learning in the school system (Erstad, 2010). This vision includes establishing transparent and consistent expectations for process and procedures around device use, encouraging integration and collaboration amongst teachers, and involving staff in the decision-making process (Danielsen 2009; Shapley, Sheehan, Maloney, & Caranikas-Walker, 2009). Rather than a more traditional leader–servant model, where the school leader takes an authoritarian approach, when setting up mobile learning policy with teachers, a person-centered leadership approach is recommended (Hadjithoma-Garstka, 2011). A delegated or person-centric approach includes checks and balances to the teachers and educational community involved in the mobile learning roll out. Prior to handing out mobile technologies to teachers or students, school leaders ought to work closely with teachers to help them plan, enact and develop a strategy to continuously improve their technology literacy skills (Hadjithoma-Garskta, 2011; Keane & Keane, 2017; Smith, 2008). The leadership should encourage and help develop a professional learning community where administrators and teachers are meeting on a consistent basis, evaluating the technology use, reviewing current research, and coplanning. Delegated leadership allows the school leaders sets the pace and provides direction, but also creates a supportive community of learners has been found to be effective (Keane & Keane, 2017).

Administrators should create professional development communities that allow teachers to collaborate and work together (Hadjithoma-Garskta, 2011; Keane & Keane, 2017). Professional learning communities can support the leadership to make informed decisions on mobile technology policies for classroom instruction. The goal should be to create policies that are flexible enough to allow instructors to personalize choices of how and when the devices are used in order to meet the learner's needs. At the same time they need to be rigorous enough that the school can make certain it protects student data and privacy rights. If the structures are too rigid, then instructors may not be able to make choices allowing differentiation for their individual classroom settings, thus taking away much of the power of using mobile technologies. A delegated approach can allow opportunities for all the stakeholders to play a role in the decision-making around the tools selected and policies created (Tedre et al., 2011).

When selecting devices, and when developing responsible use policies around mobile devices, it is imperative that the stakeholders in the educational community are supportive and involved in the decision-making process (O'Donovan, 2009). These stakeholders may include teachers, students, parents, technicians, vendors and administrators. Without opportunities for the people who are implementing the mobile devices and working in the program every day, the program is at risk for failure (Johnson & Maddux, 2008). Baltimore County Public Schools in Maryland is an example of a school district that has a comprehensive and strongly delegated leadership approach to their one-to-one mobile learning program (The Baltimore County Public Schools Technology Plan, 2014). Baltimore County Public Schools developed a digital learning committee and strategic plan committee that included stakeholders from without and outside of the school system in the decision-making and education around their one-to-one mobile device program. The committees worked together to develop a comprehensive long-term mobile learning plan to support the one-to-one initiative. By including stakeholders from a wide variety of perspectives, it allowed them to address and make informed decisions about the numerous complexities that come with a mobile learning program.

PROFESSIONAL DEVELOPMENT

Studies on one-to-one programs have shown that the teacher plays a significant role in whether or not a mobile device program is successful (Bebell & Kay, 2010; Bebell & O'Dwyer, 2010; Penuel, 2006; Sharply et al., 2009). In fact, Keane, Lang, and Pilgrim (2012) point out that well-prepared teachers did not base the success factor for a one-to-one program on the selection of the device, but rather the use of the tool. Teachers' own preferences and approaches to instruction will affect the success of the mobile learning program (Cervantes et al., 2011). Furthermore, teachers are unsure how to leverage mobile devices for learning. They need assistance in not only learning how to use the tools, but how to implement effective pedagogical approaches with the devices (Ally, Grimus, & Ebner, 2014; Hylén, 2012; Traxler, 2010). Few teachers have been trained in effective use of mobile devices for adding value to learning (Hylén, 2012). In the United States, only 7% of teachers involved in one-to-one programs received any type of pedagogical support as part of the implementation of the new program (Warschauer et al., 2012). Receiving little or no professional development in mobile learning is counterproductive to a successful implementation because any substantial technology use requires a thorough change in the view of teaching and learning, including the relation between teaching and assessment, teachers' role, and how students view and participate in school work (Rosso, 2010). Both lack of skill for troubleshooting technical problems as well as lack of know-how to properly integrate technology tools with curriculum are two big impediments for teachers in being effective with one-to-one or BYOD programs (Learning and Technology Policy Framework, 2013).

Professional development for teachers must be a priority in order for mobile learning to succeed (Kukulska-Hulme, Sharples, Milrad, Arnedillo-Sánchez, & Vavoula, 2011). According to Khaddage et al. (2015), many educational institutions have no standard policy for supporting teachers in evaluating and integrating apps for learning. Studies have also found that professional development focused on teacher's individual tech skills was not nearly as important as allowing time for teachers to collaborate and work together with the technology tools (Keane & Keane, 2017). Districts need to help teachers learn new pedagogical practices on how to use strong frameworks to implementing mobile learning tools that are based on research around effective teaching with technology (Hylén, 2012). Pedagogy should include a formalized mobile learning framework that instructors can use to plan for mobile integrated learning. One example of a framework developed to address the instructional support need for mobile learning is M-COPE (Dennen & Hao, 2014). M-COPE centers on scaffolding teachers to make effective pedagogical choices when using mobile tools.

Beyond a framework, teachers need to stay up to date on the changing landscape of how technology tools and teaching work together. Thus, schools should consider innovative approaches to long-term professional development, rather than traditional "one-and-done" models. For example, there are numerous education organizations such as the Friday Institute, eMints or Digital Promise, that are providing resources for schools and teachers to gain professional development on digital literacy through research-based microcredentials and open-source MOOCs (Acree, 2016; Making Professional Learning Count, 2015). Collaboration is a key element to successful professional development. Teachers should work together in planning and assessing their lessons and management ideas (Greaves,

Hayes, Wilson, Gielniak, & Peterson, 2010; Keane & Keane, 2017). Strong leadership can provide time and space for creating collaborative learning communities among teachers. Schools with mobile learning programs can develop collaborative professional development models such as: creating communities of practice among educators in the district to share ideas, using a mobile learning framework to design lessons, providing an extra support when educators are implementing a new mobile lesson, encouraging teachers to share what they are learning with a wider audience, encouraging teacher teams to collaboratively participate in sustained professional development models such as microcredentials, and providing teacher teams time during the school day to collaborate on ideas.

DEVICE POLICY

At all levels, national and local, mobile device policy needs to work together with school leadership to have a positive impact on learning (Hylén, 2012). Beyond making decisions about types of devices (see Table 2.1) and software, stakeholders should be integral in developing and annually reviewing mobile device rules and regulations for classroom use in school learning. The regulations should address concerns and find ways to alleviate them without hindering learning opportunities that could occur through the devices. Device policies have potential to both hinder and help learning opportunities through mobile programs (Fritschi & Wolf, 2012). At the local level, mobile device classroom rules are often woven into the school's responsible use policy. Saline Area Schools in Saline, Michigan, developed a 3-year district technology plan and includes a committee made up of a diverse group of stakeholders (Kellstrom, 2014). While the committee recognized the potential affordances of students being able to use their own devices in their learning, the district also took into consideration concerns that teachers had over how devices may disrupt and distract from classroom learning. As a result, Saline created a schoolwide "stoplight" approach to managing BYOD devices in all teachers' classrooms (Saline Area Schools Bring Your Own Device To School Program, 2013). Each day, teachers post a green, yellow or red dot on their door. Each color represents how the mobile devices will be used in that particular classroom. The stoplight approach gives students clear and consistent expectations for their device use as they enter every classroom each day. This structure of managing the devices affords flexibility for teachers who want to use student mobile devices in learning to be able to use them, at the same time, it allows teachers who prefer that the devices were not used to be able to keep the devices from distracting from the learning.

DIGITAL INFRASTRUCTURE

As mentioned earlier in this chapter, without strong and thoughtful infrastructure in place, mobile device programs will be in danger of failing (U.S. Department of Education, Office of Educational Technology, 2016; Ng & Nicholas, 2013). A long-term sustainability plan[2] is needed for infrastructure prior to purchasing mobile devices. School leadership should look at how they can use government or private grants to help fund the infrastructure. There are national initiatives to help fund digital infrastructure in the United States, the United Kingdom and a few other European counties. In addition, the European Commission has programs to help support and sustain infrastructure while also being mindful of protecting student data (Hylén, 2012). Some schools may want to consider connecting with and investing in a reliable vendor that will provide complete infrastructure support (Anderson, 2007). Schools may also find useful the step-by-step guide of the COSN Protecting Privacy Toolkit (Protecting Privacy Toolkit, 2016) or the Mobile Privacy Principles developed by GSMA (Mobile Privacy Principles, 2012).

Beyond these basic structures in the long-term planning, national and local governments recommend that school districts ensure that students and educators not only have adequate and reliable Internet access inside of schools, but outside as well (U.S. Department of Education, Office of Educational Technology, 2016). About 78% of Americans have access to high speed Internet at home. People living in rural areas are less likely to have Internet access when compared to people living in urban or suburban areas (File & Ryan, 2014). As of 2016, about 15% of homes in Europe do not have any type of Internet access at all (Internet Access and Use Statistics, 2016). In general, having access to Internet technology allows for benefits to the whole family including employment, social cohesion, and economic growth (Davies, 2015). Consequently, it is not enough to simply have sufficient Internet access inside of school it is imperative for school leaders to find creative ways to make sure that all students and their families have access outside of school, creating a more equitable mobile learning experience for the school community.

COMMUNITY PARTNERSHIP AND CONSIDERATIONS FOR EQUITY

Successful mobile program leadership considers how the program will impact the community outside of the school buildings. The school's mobile program plan should include both parental agreement and developing a partnership with parents (Bate, MacNish, & Males, 2012; Tedre et al., 2011). In doing so, implementation of one-to-one mobile models with parent partnerships can positively impact and lesson the digital gap in underserved socioeconomic communities (Ferrer, Belvís, & Pàmies, 2011; Harris, 2010). Thus, as the planning committee is developing the mobile device policies, issues of digital equity must be considered. Digital equity includes student access to mobile devices, Internet access, and support of technology use at home and at school. Many schools mistakenly assume that a one-to-one initiative means all students will have the same opportunities to learn via a digital device. While inside of school, this may seem to be the case, in reality students are coming from diverse backgrounds and often students have had different levels of exposure to technology tools in their everyday lives. Consequently, some will have more comfort, skills, and home support than others, which makes for an inequitable playing field inside of schools. In addition, there is an "under connection" concern with access to Internet at home. In a recent survey of lower income families in the United States conducted by the Joan Ganz Cooney Center, they found that while 91% of high poverty families did have some kind of Internet access, that access was considered under connected, meaning there were barriers to the access supporting student learning effectively (Rideout & Katz, 2016). For example, 52% of lower income families said their access was slow, 26% said many people are trying to share one computer, and 20% said their Internet access was cut off because of financial reasons (Rideout & Katz, 2016). Yet it should be known that parents of all income levels have an overwhelming positive viewpoint of technology, it is often financial reasons that are creating the digital equity-gap in school to home technology use for learning (Rideout & Katz, 2016). Therefore, mobile learning plans should not only explicitly recognize that digital inequities exist but also come up with plans to alleviate these inequities. Baltimore County Public School's technology plan explicitly states in their technology plan and vision that,

> There is still a digital divide in Baltimore County between those who have access to computers and the Internet and those who do not. In order to provide a systemwide, equitable digital environment for learning, Baltimore County Public Schools will move to a One-to-One platform—meaning that each student will have a digital learning device in his or her hands. (The Baltimore County Public Schools Technology Plan, 2014, p. 3)

In addition, Baltimore County Public Schools requires that one parent or guardian is part of the technology training each year with their child (The Baltimore County Public Schools Technology Plan, 2014). This allows students who may come from homes with little to no technology training to gain access to support in their home by training their parent or guardian to support the student's learning. Furthermore, Baltimore County Public Schools developed partnerships with local business to help provide broadband in homes that do not currently have access (The Baltimore County Public Schools Technology Plan, 2014). On a global level, United Nations Educational, Scientific and Cultural Organization (Hylén, 2012) has also encouraged community and business partnerships to help enable and sustain mobile learning programs. Governments can also play a key role in reducing taxes on mobile technology tools for learning (such as luxury taxes) and subsidizing high needs school systems (Hylén, 2012). Other ideas to create more equitable policies for mobile learning include: create mobile hotspot programs, make a local map of all the no-cost wireless Internet access locations around town, set up community or mobile labs using buses, partner with local business to provide broadband access, support and training for lower income families, provide community workshops for families.

DIGITAL CITIZENSHIP, CYBERSAFETY, AND ETIQUETTE FOR STUDENTS

Mobile learning should not be implemented without a schoolwide approach to digital citizenship (Hylén, 2012). Digital citizenship[3] curriculum should be employed systematically so that it touches every student and teacher with a clear and consistent message, in particular around etiquette and cybersafety. Similar to other elements of mobile device policy, parents should be included in the education and conversation around digital citizenship. Common policies, standards and language of digital citizenship should be developed between schools, students and parents so that there are no confusing or mixed messages (Hollandsworth, Dowdy, & Donovan, 2011). Experts have advocated for schools to take a proactive approach and teach information gathering online and digital etiquette before students create negative or inappropriate digital footprints (Hollandsworth et al., 2011). Utilizing a robust curriculum such as Common Sense Media (Scope & Sequence: Common Sense K–12 Digital Citizenship Curriculum, n.d.), the entire school community can create a clear and consistent message for students around what it means to be safe, responsible and respectful with their mobile devices.

RESPONSIBLE USE POLICIES AND STUDENT DATA PRIVACY

Responsible use policies in schools must detail how the mobile devices should be used to protect student data and privacy. In particular schools should make a concerted effort to focus on digital privacy for student data in regards to national or state law. For example, in the United States laws considered should include: Family Education Rights and Privacy Act (FERPA) and Children's Internet Protect Act (CIPA).[4] In Europe, schools should be mindful of the ePrivacy Directive and General Data Protection Regulation (GDPR). While many schools are widely taking advantage of third party cloud-based services to house student data, schools tend to rely on the contracts provided by the third-party service, rather than creating their own in order to protect student data and privacy, and this can compromise student privacy (Singer, 2015). According to California attorney Diane Savage (Abilock & Abilock, 2016), many schools do not understand that federal laws covering education and student privacy such as FERPA, are generally only enforceable against educational institutions that receive federal funds, and not directly against the vendors that actually cause the violation of sharing private student data. Therefore, it is important that schools protect themselves and do not blindly rely on vendors and other institutions to protect their student data. Furthermore, it is important that parents are not giving blind consent, instead they should be giving informed consent, where schools share details about the third-party websites and digital resources that will be part of the mobile learning program (Abilock & Abilock, 2016). School leaders should make certain that teachers know to present a rationale with specific information about their digital application choices in language that both parents and students can understand.

MOBILE PROGRAM ASSESSMENT

Program evaluation and monitoring are needed to sustain and improve the use of the mobile devices in school learning (Warschauer, Cotten, & Ames, 2011; Zucker, 2005). Without a strong plan to assess the many pieces of the mobile program, it will be at risk of failure. Yet, very few one-to-one or mobile programs have a systematic approach to evaluating the program (Tedre et al., 2011). Mobile program assessment should include: collecting and analyzing data to monitor the teachers' changing affect and attitudes toward the program and their professional development needs, the infrastructure reliability (both current and future plans) and troubleshooting needs, the classroom and school device management plans and policies, community relations concerning the program, and positive and negative learning outcomes affected by the mobile program. Baltimore County Public Schools developed two ways to measure their technology use in their mobile learning program: a technology efficiency report as well as a technology inventory report (The Baltimore County Public Schools Technology Plan, 2014). These reports help to inform how the effective the current program is and what areas need to be improved upon.

DISCUSSION AND RECOMMENDATIONS

In reviewing the studies and reports on mobile learning policy, it is clear that access to devices, proper legal policies, or infrastructure alone is not enough to sustain a successful mobile learning program. Mobile learning initiatives in formal education settings will be prone to failure without other necessary pieces in place. School systems cannot ignore the often-overlooked significance of leadership, pedagogy, equity, teacher professional development, community, digital citizenship, and program assessment as part of their planning. Ultimately, the many reasons cited for schools dropping their one-to-one programs such as lack of teacher buy-in, or technical issues, seem to be symptoms of a larger challenge: leadership and professional development. The following are two key recommendations, that formal school systems should put in place in order to effectively implement and sustain mobile learning programs. Above everything else leadership is fundamental. School leaders must create a strong vision, purpose, and person-centric approach to mobile learning that fosters professional learning communities and collaboration among teachers. The leadership creates decision-making bodies inclusive of stakeholders, both inside and outside the school building. The leadership should create a culture where teachers (and other relevant stakeholders) are integral in planning, creating policy, developing infrastructure plans, and assessing the mobile learning program. Second, schools should create or utilize a strong mobile learning framework and pedagogy as part of their teacher professional development and community outreach. This framework has to be carefully implemented into both the professional development of teachers and support for families. With these two pieces in place, strong leadership and a formalized professional development plan inclusive of a mobile learning framework, schools may be able to avoid some of the common problems with mobile learning and have a better chance at both positive learning outcomes and long-term sustainability.

It should be noted that it was difficult to locate mobile learning frameworks or pedagogy for formal school settings, thus this is an area that needs more research and meaningful frameworks that school systems can adopt. It was also difficult to find examples of school systems that had a formalized approach to evaluating their mobile program, again another vital field that needs further research. Moving forward, it is apparent that national policies and funding can and should help support and drive local decision-making.

NOTES

1. These include inadequate technology support for teachers and learners, lack of vision, insufficient leadership, inadequate professional development for teachers, absence of connection to the parents and community, lack of programmatic assessment, digital inequities, and insufficient infrastructure (Learning and Technology Policy Framework, 2013; *Trends in Digital Learning: How K12 Educators are Empowering Personalized Learning in American Schools*, 2016; *The Digital Playbook: Understanding the Spectrum of Students' Activities and Aspirations*, 2014).
2. A comprehensive infrastructure plan should include: network management and monitoring, network capacity planning, uninterrupted power supply, on-site technical support, reliable vendor to provide a systemic support, maintenance and upgrade of devices and equipment, insurance for devices, licensing fees for digital learning content, security filtering, and use of open standards to ensure interoperability with other learning networks (Ministry of Education Transformation Plan, 2013/2014; Thomas, 2016).
3. Digital citizenship encompasses nine elements: access, commerce, health and wellness, etiquette, law, communication, rights and responsibilities, literacy, and security (Bailey & Ribble, 2007).
4. Other U.S. laws include: Children's Online Privacy Protection Act (COPPA), Health Insurance Portability & Accountability Act (HIPAA), Individual Disabilities Education Act (IDEA), and Protecting Pupil Rights Amendment (PPRA).

REFERENCES

Abilock, R., & Abilock, D. (2016). I agree, but do I know? Privacy and student data. *Knowledge Quest, 44*(4), 10.

Acree, L. (2016). Seven lessons learned from implementing micro-credentials. Retrieved April 23, 2017, from Friday Institute: https://place.fi.ncsu.edu/local/courseblog/view.php?post=19

Ally, M., Grimus, M., & Ebner, M. (2014). Preparing teachers for a mobile world, to improve access to education. *Prospects, 44*(1), 43–59.

Anderson, C. J. (2007). A Delphi study of the strategies that will help superintendents overcome the barriers to the implementation of one-to-one laptop programs in K–12 schools by 2011. *ProQuest Dissertations and Theses, 253*.

Bailey, G., & Ribble, M. (2007). *Digital citizenship in schools.* Eugene, OR: International Society for Technology in Education.

Baker, A., Dede, C., & Evans, J. (2014). The 8 essentials for mobile learning success in education. *Qualcomm Wireless Research*. Retrieved April 23, 2017, from https://www.qualcomm.com/media/documents/files/the-8-essentialsfor-mobile-learning-success-in-education.pdf

Barshay, J. (2014, July 29). Why Hoboken is throwing away all of its student laptops. Retrieved April 23, 2017, from http://www.wnyc.org/story/why-hoboken-throwing-away-all-its-student-laptops/

Bate, F. G., MacNish, J., & Males, S. (2012, February). *Parent and student perceptions of the initial implementation of a 1:1 laptop program in Western Australia*. Paper presented at the International Conference on Education and Management Innovation, Singapore.

Bebell, D., & Kay, R. (2010). One to one computing: A summary of the quantitative results from the Berkshire Wireless Learning Initiative. *The Journal of Technology, Learning, and Assessment*, *9*(2), 4–54.

Bebell, D., & O'Dwyer, L. M. (2010). Educational outcomes and research from 1:1 computing settings. *Journal of Technology, Learning, and Assessment*, *9*(1), 4–15.

Blamire, R., & Colin, J. (2015, July). The school IT administrator: Analysing the profile, role and training needs of network administrators in European Schools. *European Schoolnet*. Retrieved April 23, 2017, from http://www.eun.org/c/document_library/get_file?uuid=2e2dcbda-f332-4a13-90e8-58098ac8d059&groupId=43887

The Baltimore County Public Schools Technology Plan. (2014, March). Retrieved April 23, 2017, from Baltimore County Public Schools Department of Information Technology website: https://www.bcps.org/offices/dot/downloads/TechnologyPlan.pdf

Cervantes, R., Warschauer, M., Nardi, B., & Sambasivan, N. (2011, May). Infrastructures for low-cost laptop use in Mexican schools. In *Proceedings of the SIGCHI conference on human factors in computing systems* (pp. 945–954). New York, NY: ACM.

Crompton, H. (2013). A historical overview of mobile learning: Toward learner-centered education. In Z. L. Berge & L. Y. Muilenburg (Eds.), *Handbook of mobile learning* (pp. 3–14). New York, NY: Routledge.

Danielsen, J. E. (2009). *A case study of one-to-one laptop initiatives in Midwest public high schools*. Available from ProQuest Digital Dissertations.

Davies, R. (2015). Broadband infrastructure: Supporting the digital economy in the European Union. Retrieved April 23, 2017, from European Parliamentary Research Services: http://www.europarl.europa.eu/RegData/etudes/IDAN/2015/565891/EPRS_IDA(2015)565891_EN.pdf

Dennen, V. P., & Hao, S. (2014). Intentionally mobile pedagogy: The M-COPE framework for mobile learning in higher education. *Technology, Pedagogy and Education*, *23*(3), 397–419.

Digital Equity Action Toolkit. (2016, February). Retrieved April 23, 2017, from Consortium for School Networking website: http://cosn.org/focus-areas/leadership-vision/digital-equity-action-agenda

The Digital Playbook: Understanding the spectrum of students' activities and aspirations. Speak Up 2013 Survey. (2014, April). Retrieved April 23, 2017, from Project Tomorrow website: http://www.tomorrow.org/speakup/SU13DigitalLearningPlaybook_StudentReport.html

Erstad, O. (2010). Addressing the complexity of impact—A multilevel approach towards ICT in education. *Assessing the effects of ICT in education*, 21.

Erstad, O. (2009): Addressing the complexity of impact—A multilevel approach towards ICT in education. In F. Scheuermann & F. Pedro (Eds.), *Assessing the effects of ICT in education: Indicators, criteria and benchmarks for international comparisons* (pp. 21-38). Luxembourg: Publications Office of the European Union.

Falloon, G. (2015). What's the difference? Learning collaboratively using iPads in conventional classrooms. *Computers & Education, 84*, 62–77.

Ferrer, F., Belvís, E., & Pàmies, J. (2011). Tablet PCs, academic results and educational inequalities. *Computers & Education, 56*(1), 280–288.

File, T., & Ryan, C. (2014). Computer and Internet Use in the United States: 2013. Retrieved April 23, 2017, from U.S. Department of Commerce website: https://www.census.gov/content/dam/Census/library/publications/2014/acs/acs-28.pdf

Fried, C. B. (2008). In-class laptop use and its effects on student learning. *Computers & Education, 50*(3), 906–914.

Fritschi, J., & Wolf, MA. (2012). Mobile learning for teachers in North America: Exploring the potential of mobile technologies to support teachers and improve practice. Retrieved from http://unesdoc.unesco.org/images/0021/002160/216084E.pdf

The Future of Mobile Learning. Implications for Policy Makers and Planners. (2013). Retrieved April 23, 2017, from UNESCO website: http://unesdoc.unesco.org/images/0021/002196/219637E.pdf

Graves, T., Hayes, J., Wilson, L., Gielniak, M., & Peterson, E. (2010). Project RED key findings. Retrieved April 23, 2017, from http://www.kent.k12.wa.us/cms/lib/WA01001454/Centricity/Domain/567/Project_RED_Presentaton_121010.pdf

Hadjithoma-Garstka, C. (2011). The role of the principal's leadership style in the implementation of ICT policy. *British Journal of Educational Technology, 42*(2), 311–326.

Harris, M. J. (2010). *Impactful student learning outcomes of one-to-one student laptop programs in low socioeconomic schools* (Doctoral dissertation), San Francisco State University.

Holcomb, L. B. (2009). Results & lessons learned from 1:1 laptop initiatives: A collective review. *TechTrends: Linking Research & Practice to Improve Learning, 53*(6), 49–55.

Hollandsworth, R., Dowdy, L., & Donovan, J. (2011). Digital citizenship in K–12: It takes a village. *TechTrends, 55*(4), 37–47.

Hu, W. (2007, May 4). Seeing no progress, some schools drop laptops. Retrieved April 23, 2017, from http://www.nytimes.com/2007/05/04/education/04laptop.html

Hylén, J. (2012). Turning on mobile learning in Europe: Illustrative initiatives and policy implications. *Teacher Focus, UNESCO Working Paper Series on Mobile Learning, Paris, France*. Retrieved April 23, 2017, from http://www.unesco.org/

new/en/unesco/themes/icts/m4ed/mobile-learning-resources/unescomobilelearningseries/

Internet access and use statistics—Households and individuals. (2016). Retrieved April 23, 2017, from http://ec.europa.eu/eurostat/statistics-explained/index.php/Internet_access_and_use_statistics_-_households_and_individuals

Johnson, D. L., & Maddux, C. D. (2008). Introduction: Effectiveness of information technology in education. *Computers in the Schools, 24*(3), 1–6.

Keane, T., & Keane, W. (2017). Achievements and challenges: Implementing a 1:1 program in a secondary school. *Education and Information Technologies, 22*(3), 1025–1041.

Keane, T., Lang, C., & Pilgrim, C. (2013). Pedagogy! iPadology! Netbookology! Learning with mobile devices. *Australian Educational Computing, 27*(2), 29–33.

Kellstrom, H. (2014, July 1). Saline area schools: District technology plan. Retrieved April 23, 2017, from https://www.salineschools.org/downloads/technology_files/final_sastechplan2014-17.pdf

Khaddage, F., Christensen, R., Lai, W., Knezek, G., Norris, C., & Soloway, E. (2015). A model driven framework to address challenges in a mobile learning environment. *Education and Information Technologies, 20*(4), 625–640.

Khan, A. I., Al-Shihi, H. I., Al-Khanjari, Z. A., & Sarrab, M. K. (2015). Mobile learning (M-Learning) adoption in the Middle East: Lessons learned from the educationally advanced countries. *Telematics and Informatics, 32*(4), 909–920.

Koole, M. L. (2009). A model for framing mobile learning. *Mobile Learning: Transforming the Delivery of Education and Training, 1*(2), 25–47.

Kukulska-Hulme, A., Sharples, M., Milrad, M., Arnedillo-Sánchez, I., & Vavoula, G. (2011). The genesis and development of mobile learning in Europe. In D. Parsons (Ed.), *Combining E-learning and m-learning: New applications of blended resources* (pp. 151–177). Hershey, PA: IGI Global.

Learning and Technology Policy Framework. (2013). Retrieved April 23, 2017, from School Technology Branch, Alberta Education website: https://education.alberta.ca/media/1046/learning-and-technology-policy-framework-web.pdf

Lemagie, S. (2010, November 21). 1 student, 1 laptop proves costly. Retrieved April 23, 2017, from http://www.startribune.com/local/109779099.html

Making Professional Learning Count: Recognizing Educators' Skills with Micro-credentials. (2015). Retrieved April 23, 2017, from Grunwald Associates LLC and Digital Promise website: http://digitalpromise.org/wp-content/uploads/2014/05/making_professional_learning_count.pdf

McLean, K. J. (2016). The implementation of bring your own device (BYOD) in primary [elementary] schools. *Frontiers in Psychology, 7.*

McClean, K. J. (2016). The implementation of bring your own device (BYOD) in primary [elementary] schools. *Frontiers in Psychology, 7,* 1–3. Retrieved from https://www.ncbi.nlm.nih.gov/pmc/articles/PMC5108779/

Ministry of Education Transformation Plan. (2013/2014). Retrieved April 23, 2017, from British Columbia Ministry of Education website: http://www.bced.gov.bc.ca/pubs/tt-plan/2013-2014_bced_tt_plan.pdf

Mobile Privacy Principles. (2012). Document: Promoting a user-centric privacy framework for the mobile ecosystem, Version 1.0. (2012). Retrieved April 23, 2017, from GSMA website: http://www.gsma.com/publicpolicy/wp-content/uploads/2012/03/gsmaprivacyprinciples2012.pdf

Muir, D. (2013, March). International perspective on mobile learning. *Society for Information Technology & Teacher Education International Conference, 1*, 819–843).

Murray, O. T., & Olcese, N. R. (2011). Teaching and learning with iPads, ready or not? *TechTrends*, *55*(6), 42–48.

Ng, W., & Nicholas, H. (2013). A framework for sustainable mobile learning in schools. *British Journal of Educational Technology, 44*(5), 695–715.

O'Donovan, E. (2009). Are one to one laptop programs worth the investment? *District Administration.* Retrieved November 17, 2017, from https://www.district-administration.com/article/are-one-one-laptop-programs-worth-investment

Penuel, W. R. (2006). Implementation and effects of one-to-one computing initiatives: A research synthesis. *Journal of Research on Technology in Education*, *38*(3), 329–348.

Policy Guidelines for Mobile Learning. (2013). Retrieved April 23, 2017, from UNESCO website: http://unesdoc.unesco.org/images/0021/002196/219641e.pdf

Protecting Privacy Toolkit. (2016). Retrieved April 23, 2017, from Consortium for School Networking website: http://www.cosn.org/focus-areas/leadership-vision/protecting-privacy

Quinn, C. (2000, Fall). MLearning: Mobile, wireless, in-your-pocket learning. Retrieved April 23, 2017, from http://www.linezine.com/2.1/features/cqmmwiyp.htm

Rideout, V., & Katz, V. (2016). *Opportunity for all? Technology and learning in lower-income families.* New York, NY: The Joan Ganz Cooney Center at Sesame Workshop.

Rosso, J. (2010). Disruption in the educational paradigm: Notes on 1:1 research. Research watch, K–12 Blueprint. Retrieved from http://www.k12blueprint.com/content/disruption-educational-paradigm-notes-11-research

Saline Area Schools Bring Your Own Device To School Program. (2013, February 25). Retrieved April 23, 2017, from Saline Area Schools Instructional Technology website: https://docs.google.com/file/d/0B1ehprvXTUsVQnFuUzNvRGhWczQ/edit

Scope & Sequence: Common Sense K-12 Digital Citizenship Curriculum. (n.d.). Retrieved April 23, 2017, from https://www.commonsensemedia.org/educators/scope-and-sequence

Shapley, K., Sheehan, D., Sturges, K., Caranikas-Walker, F., Huntsberger, B., & Maloney, C. (2009). Evaluation of the Texas Technology Immersion Pilot: Final outcomes for a four-year study (2004–05 to 2007–08). Austin, TX: Texas Center for Educational Research.

Shuler, C. (2009). *Pockets of potential: Using mobile technologies to promote children's learning.* New York, NY: The Joan Ganz Cooney Center at Sesame Workshop.

Smith, D. (2008). Why do most IT projects fail? It's not because of the technology. *Portland Business Journal.* Retrieved November 15, 2017, from http://www.bizjournals.com/portland/stories/2008/10/20/smallb4.html?page=all

Singer, N. (2015, March 2). Privacy pitfalls as education apps spread Haphazardly. *New York Times*. Retrieved April 7, 2015, from http://nyti.ms/19csX4Z

Tedre, M., Hansson, H., Mozelius, P., & Lind, S. (2011). Crucial considerations in one-to-one computing in developing countries. In P. Cunningham & M. Cunningham (Eds.), *Proceedings of IST-Africa 2011 Conference* (pp. 1–11). Gaborone, Botswana: IST-Africa Initiative.

Thomas, S. (2016). Future ready learning: Reimagining the role of technology in education. 2016 National Education Technology Plan. Washington, DC: Office of Educational Technology, U.S. Department of Education.

Traxler, J. (2005, June). Defining mobile learning. In Proceeding of *IADIS International Conference Mobile Learning, 2005*(pp. 261–266).

Traxler, J. (2010). Sustaining mobile learning and its institutions. *International Journal of Mobile and Blended Learning, 2*(4), 58–65.

Trends in Digital Learning: How K12 Educators are Empowering Personalized Learning in American Schools. (2016, June). Retrieved April 23, 2017, from Project Tomorrow website: http://bbbb.blackboard.com/2016digitaltrendsreport-ty

U.S. Department of Education, Office of Educational Technology. (2016). Future ready learning: Reimagining the role of technology in education: 2016 National Education Technology Plan. Retrieved from http://tech.ed.gov/files/2015/12/NETP16.pdf

Valiente, O. (2010). *1-1 in education: Current practices, international comparative research evidence and policy implications* (OECD Education Working Papers, No. 44). Paris, France: OECD.

Warschauer, M. (2006). *Laptops and literacy: Learning in the wireless classroom.* New York, NY: Teachers College Press.

Warschauer, M., Cotten, S. R., & Ames, M. G. (2011). One laptop per child Birmingham: Case study of a radical experiment. *International Journal of Learning and Media, 3*(2), 61–76.

Zucker, A. (2005). Starting school laptop programs: Lessons learned. *One-to-One Computing Evaluation Consortium, 1*, 1–6.

CHAPTER 3

MOBILE LEARNING CURRICULA

Policy and Potential

David Parsons
The Mind Lab by Unitec

Kathryn MacCallum
School of Computing, Eastern Institute of Technology

This chapter explores the relationship between mobile learning and policy as it pertains to the curriculum. Instructional content is often guided by national or international curricula, which reflect the education policies of government agencies or other organizations. Few of these curricula directly address mobile learning, but many include reference to mobile digital tools, and many of their themes resonate with the qualities of learning in a mobile age. Does this mean that current curricula are adequate for effective integration of mobile digital tools? We would argue that they are not, since their integration of mobile learning is mostly implicit, sporadic and unfocused. To address these shortcomings, we have constructed a new curriculum model that incorporates mobile affordances, the journey to digital fluency and applied learning activities into

Mobile Learning: Perspectives on Practice and Policy
pp. 45–64

a layered view of what we believe a mobile learning curriculum should encompass.

Existing definitions of mobile learning in the context of curriculum are problematic from a number of perspectives. It has been suggested in an earlier study by Botha, Batchelo, Traxler, de Waard, and Herselman (2012) that a mobile learning curriculum would comprise three broad learning objectives; knowledge of mobile learning itself, facilitation of mobile learning, and understanding its implications. These are relevant goals for some contexts, but they are too narrow for addressing the scope of mobility in the broader digital curriculum. New definitions of learning mobility, and 21st century styles of teaching and learning, demand a somewhat more ambitious reading of how learning with mobile devices fits into a contemporary curriculum. Curricula need to be aligned effectively towards the kinds of learning outlined by recent research into transferable competencies, such as the innovative teaching and learning (ITL) rubrics that identify six areas of 21st century skill (ITL Research, 2012), all of which may to some extent be supported by mobile learning. In this chapter, we provide some insights into how curricula can be designed to take full advantage of learning with mobile devices, not just with studies that focus on mobile learning, or even studies that focus only on ICT skills (just one of the 21st century areas of skill) but throughout the curriculum. It is hoped that such insights will assist policymakers in considering future curriculum changes. Curriculum design is not neutral, and makes certain assumptions about the type of education that is required and the ways that it may be taught. Examining present and future curricula through the lens of mobile learning allows us to consider whether traditional views of curriculum design are still appropriate in an era of pervasive digital tools.

THE MOBILE DIGITAL CLASSROOM

The adoption of BYOD (Bring Your Own Device) policies in education, and the increasing focus on one-to-one (1:1) computing, where students are exposed to portable digital devices in the classroom, has begun to impact on education systems worldwide (Male & Burden, 2014). This technology revolution means that student access to computers is less likely to be limited by restricted computer labs or sharing a single classroom computer between many students. Those fortunate enough to have access to their own digital devices are now able to bring them into learning spaces and are increasingly encouraged to utilize them to develop, create and communicate. In many countries, the aim to give every student access to a device has been a major focus across a range of institutions and sectors. Schools have initiated projects where students purchase, or are given or loaned, a range of portable digital devices that they can use in their day to day learning (Male & Burden, 2014). These initiatives have either been driven by the school management, their boards of trustees, the teachers or students themselves or, as in Australia, the government. In that case, the Digital Education Revolution national initiative was undertaken between 2009 and 2013 to provide laptops to all Year 9–12 students (Buchanan, 2011). Such national policies are, however, rare. Cutting across international boundaries there is the One Laptop Per Child initiative (OLPC, 2016), which aims to provide each child in the program access to a computer. The program is targeted at less affluent communities, to provide them with access to a rugged, low-cost, low-power, connected device such as a tablet or laptop.

Defining the Role of Mobility Within Education

Underpinning the systematic way that mobile technology has being brought into the contemporary classroom, there is an acknowledgment that digital skills and concepts are vital for 21st century learning. A number of countries have started to introduce digital skills into their core curricula (Ananiadou & Claro, 2009). When these are examined, it is reasonable to assume that mobile learning may support some of their objectives. The difficulty that many educators seem to have when attempting to define a mobile learning curriculum is in successfully identifying curriculum aspects that are indeed mobile, rather than merely digital (Botha, Batchelor, Traxler, de Waard, & Herselman, 2012). Over recent years, as the range of smaller digital devices has increased markedly, the definition of a mobile device has become less clear than it once was. In addition, as these common devices become more and more integrated into everyday classroom practice, for example with the increasing popularity of the BYOD classroom, the concept of identifying a separate mobile curriculum may appear to be increasingly irrelevant and anachronistic. However, we should consider that there are some specific and important affordances of mobility that can be identified and, further, related directly to certain types of curriculum practice. If we want to develop a viable concept of a mobile learning curriculum, we need to begin with some analysis of affordances that makes it meaningful to refer to mobile digital devices, rather than just learning with digital devices, whether mobile or not. Once we understand what it is that can be achieved in teaching and learning with mobile devices, and indeed with mobility itself, then we might usefully be able to encapsulate the core skills that might be embedded in such a curriculum.

Exploring Mobile Learning Affordances

Bower (2008) noted the importance of perceived affordances and highlights the three-way relationship between technological affordances, social affordances and educational affordances. With the increasing use of mobile social media and collaboration in the 21st century classroom, these issues seem particularly appropriate when considering mobile learning curricula. Bower outlined eleven e-learning technology affordances but, of these, only the temporal affordance of access anytime anywhere (accessibility), and synthesis affordances (the capacity to combine multiple tools together to create a mixed media learning environment), resonate particularly with mobile learning. In a similar vein, Conole and Dyke (2004) described ten affordances of ICT. Again, there are few direct links with mobility, though under "diversity" the authors suggested a role for virtual (and perhaps augmented?) reality, increasingly a feature of mobile and wearable devices. The "multimodal and non-linear" affordance may also offer insights into the value of mobile learning; "the experimental and pragmatic way in which most people acquire for themselves skills in ICT" (p. 119). Brown (2015) provided another list of 16 affordances. She made the point that many analyses of affordances are considered at a high level of abstraction, whereas in her study all the affordances were identified at a much more detailed level, explicitly linked to a study around the use of ICT tools in mathematics. From these various authors, it becomes clear that there is no consistent list of affordances for learning with technology, and that any use of affordances as a tool for designing curricula needs to be targeted at an appropriate level of abstraction and contextualized to a particular learning environment.

Given that generic analyses of affordances in learning technologies do not provide much specificity regarding mobile learning, we need to turn our attention to literature that has explicitly addressed affordances in that area. Ciampa (2014) referred to the motivational affordances of mobile devices, applying Malone and Lepper's (1987) taxonomy of challenge, curiosity, control, recognition, competition and cooperation to a mobile learning context. While this taxonomy provided a good match for the motivational experiences of the students using mobile devices in the reported study, particularly in terms of cooperation, no specifically new mobile learning affordances were identified. From another perspective, affordances have been characterized by how they are used. Naismith Lonsdale, Vavoula, and Sharples (2004) classified the use of mobile devices along the two dimensions of personal versus shared (with an implicit impact of collaborative activity) and portable versus static (whether a device that can be used in a mobile context is, in fact being used in a static context). Lai, Yang, Chen, Ho, and Chan, (2007) stated that the affordance of mobile devices is underpinned by its ability to support "real-time information whenever and wherever learners need it" (p. 5). Therefore, the key characteristics of portability and immediacy are important when exploring mobile affordances. Parsons, Thomas, and Wishart (2016) further explored this approach. In their article, they identified six specific mobile affordances from the literature which differentiate mobile digital learning from other types of digital learning. These affordances are: (1) Portability: enabling a learner to move around and engage with their environment, (2) Evidence and data: to gather a rich source of information in multiple ways and share, manage or store information to support the learning process, (3) Communication: to support the engagement and interaction with others in the same environment or to a remote audience, (4) Interaction with the interface: to visualize and present digital content in a multitude of ways, (5) Contextual, active learning: for learning to be centered within the learning activity and learners to be engaged and supported by the device and support mobile inquiry, and (6) Outdoor environment: To support learning outside the classroom in formal and informal activities. Two further affordances were identified in the article; digital creativity and remote control of other devices (such as robots or printers).

Before leaving the concept of affordances, it is worth taking note of some interesting ideas put forward by Scarantino (2003). First, that there are two classes of affordances; goal affordances (where something is done) and happening affordances (where something happens). Further, he noted that the probability of perceiving an affordance may vary depending on current knowledge. This raises interesting questions, first around how to scaffold mobile learning such that potential affordances are perceived by the learner based on their current state of mobile digital skills, and also how these skills in turn mediate these perceptions.

DIGITAL PROFICIENCY, DIGITAL LITERACY, DIGITAL FLUENCY

Although a range of mobile learning affordances have been identified in the literature, for mobile tools to be effective within education both students and teachers require the necessary literacy to harness these affordances. Literacy is the ability to identify, understand, interpret, create, communicate and compute texts (written, printed, and digital) (Bawden, 2008). Digital literacy places these abilities specifically in the context of a variety of different digital sources and may be regarded simply as literacy in the digital age. The concept is wider than just technical skills. Rather, it relates to the ability to understand and use information in multiple formats from a wide variety of sources (Bawden, 2008). The English primary curriculum defines digital literacy as learners being able to express themselves and develop their ideas through ICT both for work readiness and as digital citizens (Department for Education, 2013). Park (2016) includes competency in computational thinking as part of digital literacy. Literacy, however, may not be sufficient, and digital fluency may be required to gain the fullest benefit from mobile learning. The New Zealand Ministry of Education have, among others, emphasized the importance of digital fluency, and refer to such mobile learning related issues as learning anywhere, anytime, and digital learning for every learner regardless of location (NZ MoE, n.d.) What, though, is meant by digital fluency? Wenmouth (as cited in Spencer, 2015) describes it as being the third stage of showing wisdom and confidence in the application and use of digital technologies, building on both digital proficiency and digital literacy. Digital proficiency is seen here simply as having a set of skills in using digital tools, while digital literacy requires the application of those skills and "thinking digitally." Digital fluency, in this model, requires digital citizenship. It is suggested that digitally fluent learners can self-select from a range of tools to achieve outcomes, and navigate collaborative spaces effectively and confidently (Spencer, 2015). Park (2016) further expands digital fluency to include "digital intelligence." According to Park, digital intelligence is the set of social, emotional and cognitive abilities that enable individuals to face the challenges and adapt to the demands of digital life. This can be defined within three levels: digital citizenship, digital creativity and digital entrepreneurship. Of these, the author suggests that digital citizenship is the level that is least addressed in current curricula.

Further to this concept of digital fluency (or intelligence) is the ability of students to take a wider, more encompassing view of digital tools, and apply them to analysis and problem solving. Wider concepts of systems, design and computational thinking have been identified as important themes that support digital fluency. Systems thinking is the ability for individuals to understand the holistic approach needed to identify and solve problems that impact and are influenced by a set of interrelated and complex systems. Design thinking is a set of strategies that are needed to understand, ideate and design solutions. These solutions require the ability to generate creative and innovative ideas, also to plan, execute, analyze and evaluate these ideas to ensure success. Computational thinking also focuses on problem-solving but is rather "the method that is applied to create solutions that can be implemented using digital technologies. It involves integrating strategies, such as organizing data logically, breaking down problems into parts, interpreting patterns and models and designing and implementing algorithms" (ACARA, n.d.). These specific features have been identified as important concepts when developing digital fluency in the curriculum, in particular the Australian Digital Curriculum uses these three constructs to underpin their digital curriculum framework.

Mobile Digital Fluency

Some researchers have attempted to locate digital literacies and associated concepts within mobile technology (Havelka, 2013; Parry, 2011). One example is the development of the Mobile Digital Literacy Framework (Ng, 2013). This framework has three dimensions: technical (and functional), cognitive (critical thinking, evaluative and multiliteracies capabilities, and competencies) and social-emotional (attitudes, social skills). These three dimensions interact and develop multiple literacies that need to be cultivated to encourage a mobile digitally literate individual. It is notable, however, that the specifics of mobility barely surface in the framework, except with reference to their limitations. One might make similar observations about Pachler, Cook, and Bachmair's (2010) socio-cultural ecology of mobile learning, where the concepts of structures, agency and cultural practices are not explicitly focused on the features of mobility. This is not to say that such analyses do not have insights to offer about mobile learning. Rather that, if we are to surface the unique features of what might be a mobile learning curriculum, we will need to try to see beyond generic concepts.

As a result of the above discussion, it is clear that digital fluency plays a significant role in the effective use of technology. Given that the spectrum of digital competence, literacy, and fluency is both complex and contested, we must acknowledge that the definitions presented in this chapter are not universally agreed upon. Further, when examining the specifics of mobile learning, the question of whether this requires a different set of skills or a subset of broader digital fluency is still to be debated. To contribute to this debate, in the next section we explore to what extent these concepts are acknowledged within existing curricula.

EXPLORING THE DIGITAL CURRICULUM

Internationally, educators have begun to acknowledge that digital fluency is as important a skill as traditional literacy and numeracy, and therefore have started to address these needs within the curriculum. Concepts such as computational thinking, systems thinking and design thinking have been seen as important mind-sets that underpin the ability of students to better engage with digital technologies. Many countries have started to embed digital literacy within their national curricula. For example, Australia has established the Digital Technologies Curriculum, which focuses on knowledge, understanding, and skills across a range of systems to "use computational thinking and information systems to define, design and implement digital solutions" (ACARA, 2013). The United Kingdom has embedded within their curriculum goals that all students are to understand and apply fundamental computer science principles, such as abstraction, logic, algorithms, and data representation, and write computer code to solve problems (Department for Education, 2013). In contrast, other countries, such as New Zealand, have only recently started to address how this could be embedded into the teaching curriculum (Ministry of Education, 2017). However, the current NZ curriculum includes connectivist concepts of students creating knowledge through connections in an open and flexible curriculum (Starkey, 2010). In the United States, the Common Core standards embrace English language arts and literacy in history, social studies, science, and technical subjects, as well as mathematics, but digital skills are not specifically included. It is left to third parties, such as the Gates Foundation (2011), to provide suggested mappings between the common core and digital skills.

A number of authors have undertaken studies of various curricula to determine how they address technology. Rasinen (2003) examined the technology education curricula of six different countries: Australia, England, France, The Netherlands, Sweden, and the United States. The purpose of the study was to inform the new national curriculum in Finland. In each case, the scope and focus of each curriculum element was defined (rationale, theory, objectives, methods, content, and means of evaluation). In the analysis, curricula are noted as being either "lehrplan-type," which provide specific details of what should be taught and how it should be taught, or "standards-type," which specify the goals that should be met, but do not specify the actual curriculum. The study also highlights that the Australian curriculum stresses the importance of lifelong learning and developing innovative skills. Although the analysis reported in the study does not directly refer to mobile learning, its summary of curriculum objectives, methods and contents from the perspectives of society, school and the student identifies some repeating themes that are core to cross curriculum technology: learning by doing, increased cooperation and collaboration both within and beyond school, seeing technology as part of wider society, integration across subject areas, and gender equity. All of these themes can be seen consistently in the mobile learning literature.

Another international study was undertaken by Aesaert, Vanderlinde, Tondeur, and van Braak (2013), though in this case, the national educational technology curricula of only three nations were included; Norway, Flanders, and England. The authors focused in particular on 21st century skills and digital literacy and noted that, despite the differences in detail between the national curricula, they shared the central themes of higher-order thinking skills using educational technology and information processing rather than technology use itself. Although, again, there were no specific references to mobile learning, certain curriculum examples would lend themselves to the use of mobile devices, such as using sensors to record environmental data.

MOBILE TECHNOLOGY WITHIN THE CURRICULUM

The preceding examples of curricula indicate some technology-related skills that successful learners need, but how can specifically mobile technology can be drawn into the digital curriculum? In this section, we provide some examples of how reference to mobile devices and activities has been incorporated into some existing curricula.

One of the problems that we face in seeking to identify where mobile learning fits into broader digital curricula is the way that explicit reference to mobile devices is frequently confined to technical rather than pedagogical contexts. For example, the Victorian Digital Technology Curriculum (Victoria State Government, n.d.) only refers to mobiles in the context of wireless networking, thus any linkage with broader curriculum aims are implicit. We might identify wireless-supported activities such as creatively communicating ideas in a range of settings, and communicating and collaborating with known and unknown audiences as strengths of mobile tools and channels, but the curriculum does not do so. Some curricula, however, offer enough links to mobility to draw out some more explicit ideas.

The Australian Technology Curriculum (ACARA, 2013) provides some ideas that acknowledge the role of mobile devices in broader pedagogy. For example, it refers to the way that mobile systems are transforming learning, recreational activities, home life and work, and that they have led to new ways of thinking, collaborating and communicating when developing social and intellectual capital. The role of mobile devices as problem-solving tools in a knowledge-based society is described with some examples, such as using a mobile device to record musical performances, sharing knowledge of how to share digital media and discussing policies about using mobile phones for learning and accessing social media websites at school. Another Australian document (Digital Education Advisory Group, 2013) refers to how smart mobile devices can "change the game," citing an example of how history students studying the Vietnam War used mobile devices to record interviews with veterans and edit them together with other materials to share online.

The English National Curriculum focuses on teaching students how to think creatively (Department for Education, 2013). The use of mobile technology is seen as a way to enable students to explore and present their work in more dynamic and creative ways (Leask & Meadows, 2000).

In addition to these government-led initiatives, the Association for Computing Machinery (ACM) provides some further examples of how mobile learning can be embedded into the curriculum. The ACM provides undergraduate curriculum guidelines for five subdisciplines; computer science, computer engineering, information systems, information technology and software engineering. It should be noted that these curricula are intended for computing programs rather than for broader provision of computer-related skills across other disciplines. Thus, they address a different purpose than some of the other curricula described in this chapter and are, of course, targeted at undergraduate students rather than school students. In addition, however, the Computer Science Teachers Association (part of the ACM) publishes the K–12 Computer Science Standards for schools (CSTA Standards Task Force, 2011). Although these are again focused on computer science teaching rather than generic use of mobiles for learning, the document considers some important pedagogical issues. The various ACM undergraduate curricula have been compiled at different times by different groups of academics, so it can be informative to look across these curricula for ideas about how mobile learning has been, or could be, integrated. None of them explicitly mention mobile learning, but all refer to some degree or other to the role of mobile devices, and some give important pointers to their potential uses in education.

The Computer Science Curriculum (ACM, 2013) acknowledges not only that students can learn about mobile environments but that they can also learn within them. It also notes the way that mobile devices can impact on social interactions as part of the social context topic and touches on mobile augmented reality. The Computer Engineering Curriculum (ACM, 2004) notes that computers, including mobile phones, have replaced more conventional electronic devices, highlighting special characteristics of mobile systems such as location awareness. The Information Systems Curriculum (ACM, 2010) references ubiquitous mobile computing and the way that it has impacted on organizations and individuals worldwide but, perhaps surprisingly, goes no further. The Information Technology Curriculum (ACM, 2008) refers to the way that mobile technology means that communications and connectivity are expected anytime and anyplace, one of the oft-mentioned characteristics of mobile learning. It also highlights the differences between mobile and wearable devices, and highlights mobile spatial databases as a curriculum topic.

The least relevant curriculum is that in Software Engineering (ACM, 2014), which only references mobile in architectural terms, not in usage.

In contrast, the K–12 Computer Science Standards provide more of a pedagogical focus (CSTA Standards Task Force, 2011). They note how elementary school students can work cooperatively with other students and teachers by using mobile technology, gathering and communicating information. The standards also recognize that students can use mobile devices for problem-solving and self-directed learning. Being a computer science curriculum it also encourages the design and development of software products such as mobile applications, in the context of demonstrating and communicating curriculum concepts.

The UNESCO policy guidelines for mobile learning (UNESCO, 2013), while not directly concerned with curricula, make some important points about the relationship between the curriculum and mobile devices. They note that mobile tools allow for immediate feedback and assessment, allowing for a more flexible approach to curricula than the traditional model of delivering content and following up with summative assessment. Mobile applications that provide both learning content and instant feedback can shift the emphasis of a curriculum from content and measurement to active, personalized learning, freeing educators to spend more time working directly with students. The guidelines also refer to the curriculum in teacher training, and the way that mobile devices can be used to spend more time practicing pedagogy in authentic environments. However, this concept of moving the curriculum into authentic environments through mobility is by no means limited to teacher training.

The above examples explain how the curriculum can be enhanced using mobile technology, but there are also examples of mobile specific curricula. The most notable is that suggested by Botha et al. (2012), where they proposed a mobile learning curriculum framework that addresses the process of developing a curriculum, some evaluation criteria (relevance, consistency, practicality and effectiveness) and an overarching organization into themes and modules. The framework design is underpinned by three broad learning objectives; acquiring domain knowledge, developing skills for mobile learning practice and understanding the relationship between domain knowledge and application contexts. One of the key assumptions of the article is that "mobile technology includes the capability to extend the geographical and social reach of education systems and opportunities to enrich and enhance the nature of education" (p. 5) such that mobile learning is part of a wider social use of mobile devices. The five themes in the curriculum are: the impact of mobiles on people, communities and societies, the impact of mobiles on the economy, the impact of mobiles on learning, the nature of the technology, systems and devices, and becoming mobile. Although this is a relevant view of mobile learning curricula, one might regard it as being somewhat siloed into the domains, perhaps at the expense of the skills for practice. Further, what is not clear from such a thematic approach is how digital fluency is to be fostered across learning contexts.

Based on the above examples, we can see that mobile technology has the potential to be a strong force within curricula for supporting and engaging learners. There are, however, still gaps with these existing approaches. The following section addresses some of the challenges of developing a mobile specific curriculum, before we introduce a potential structure that could be used to develop and integrate mobility and mobile technology into curricula at all levels.

The Challenges of Developing a Mobile Curriculum

While several curricula acknowledge in a fragmentary manner the importance of mobile tools in contemporary learning, a wider view is needed, specifically around how teachers can leverage mobile environments. Pegrum, Oakley, and Faulkner (2013) reported that the strongest point to emerge from teacher interviews in their study of ten schools in Western Australia was the need for professional development on how to integrate m-learning with the curriculum and pedagogy. Chen and Denoyelles (2013) found that there was a distinct gap between students that own mobile devices and those that use them to support their learning. The reason for this disparity was that educators did not have the necessary know-how and appropriate pedagogy to effectively embed mobile technology into the curriculum, and that these problems were often exacerbated by issues of inequitable access and supporting multiple types of devices. Looi et al. (2014) pointed out that it is rare to see mobile learning innovations become immersed in the curriculum or an integral part of routine classroom practices. To address this limitation, the authors report on their work to integrate mobile learning into the science curriculum of Singapore, using the "Mobilized 5E" model of engagement, exploration, explanation, elaboration and evaluation. Their new vision of the curriculum is inquiry-based and leverages the affordances of mobile technologies. One interesting feature of their work is that the mobile learning curriculum was implemented through an integrated set of mutually supporting mobile tools that together enabled the students to independently perform their science inquiry and take ownership of their learning, the "synthesis affordance" introduced earlier in this chapter. A particular characteristic that marked this learning process out from more traditional methods was the involvement of mobile activities outside the classroom. The authors point out that the success of the curriculum innovation was due to the methodologies employed in curriculum development and strategies that supported teachers in implementing that curriculum, whereby a continuous consultative process is required that both analyses the past and looks to the future.

Further to the challenges above, curriculum design in any technology related area at a given point of time may be outdated within a few years due to the rapid and sustained transformation brought on by pervasive digital technology (Eaton, 2015). This perhaps emphasizes the need to frame curricula not in subject and content terms but rather in terms of 21st century skills. ITL Research (2012) describe six rubrics for 21st century learning: collaboration, knowledge construction, self-regulation, real-world problem-solving and innovation, the use of ICT for learning and skilled communication. Unlike specific content, skill based curriculum design will remain relevant over a long period of time. One question that arises is the role of computational thinking and coding in the curriculum (Eaton, 2015). As with other technologies, it is important to be able to extract generic skills from the use of specific coding tools that may quickly become obsolete. It would seem therefore that the curriculum should not focus on coding specifically but rather on broader aspects of digital creativity around building mobile applications. The ITL 21st century rubrics give such an example, whereby students might design and program a mobile smartphone app that could help senior citizens in their daily lives. Regardless of the tool used, the students would build knowledge beyond the technology to consider their users and their social context. At the time of writing, there are a number of different types of application construction tool that would allow focus on design and context without being too focused on coding. Mobile Android apps can, for example, be built visually in MIT App Inventor, while mobile games can be built on the iPad with tools like Hopscotch. HTML5 applications for mobile devices can be built interactively using cloud-based tools such as Appery, while location based, gamified, mobile learning experiences can be created with tools such as ActionBound. Tools like these allow students to become creative mobile application builders with a high level of productivity, enabling them to focus more on iterative analysis of user requirements, design, and testing. The same skills remain important even as these specific tools become superseded by others. It should be noted that some tools allow mobile applications to be created using mobile devices, while others create mobile applications from the desktop for deployment to mobile devices on completion. In either case, it is the higher-level skills that are being exercised, with the end focus on the value of creating a particular mobile user experience.

Earlier in this chapter we referred to the role of affordances in defining a mobile learning curriculum. Several these affordances (portability, evidence and data gathering, contextual, active learning, and outdoor environment) come together in learning activities where mobile devices are used as "probes," in the sense that learners can use them to go out into the world to collect information. The addition of multiple new sensor types has been one of the most significant changes to mobile device capability in recent years. Today's mobile device is a powerful sensing platform that enables the user to distill meaningful views of their activities (Cardone, Cirri, Corradi, Foschini, & Maio, 2013). There are three generic types of sensor fitted to mobile devices; motion, environment and position (Android Developers, n.d.) Position sensors can be used to guide learners to different locations, while environmental sensors allow learners to explore and measure their environment, for example, weather, noise pollution, light and shade, geography (orientation, elevation), and so forth. Motion sensors can be used for some types of measurement, and also for control of other devices, such as robots. Where sensors are not built into the devices, separate sensor hardware can be linked wirelessly for specialist tasks such as gas or chemical detection. Sensors provide extensive support for mobile inquiry learning tasks across environments. As with software tools, sensor technology will continue to evolve and provide further opportunities for mobile student inquiry in the curriculum, supporting the development of real world learning and critical thinking skills.

A MOBILE LEARNING CURRICULUM STRUCTURE

In this section, we draw together the various mobile learning curriculum sources that have been discussed during this chapter and formulate a possible structure for mobile learning curricula that takes account of foundational skills as well as higher level activities. In creating this structure, we are acknowledging that there are two strands to mobile digital curricula. One is the way of becoming a mobile learner, such that the use of mobile tools for learning is integral to the learner's thought processes. This is the journey towards digital fluency. We have based this structure on several sources that discuss these types of generic digital skills. The other strand relates to what this means in practical terms, how this mode of learning is applied in particular contexts. Therefore, in our curriculum structure we have attempted to align these two views of the curriculum into a cohesive view of what mobile learning might involve at each level. We have based this part of the structure on our literature review of a range of international curriculum documents relating to mobile learning. Running through the center of this model, and binding it together, is the set of mobile affordances. Some indicative learning activities drawn from sources cited in this chapter are also included at each level.

Figure 3.1 outlines the basic elements that we have identified as being core to a mobile learning curriculum. This model follows a ziggurat style, whereby there are a series of three levels of mobile learning that support each other. The central column contains the mobile affordances. To the left are the three steps towards digital fluency, with the overall type of learning indicated at each level. To the right are the core types of learning at each step, with some indicative learning activities. Between the three layers are the thresholds that enable learners to move to the layer above.

Unlike the themes and modules approach taken by Botha et al. (2012) we have based our structure on a foundation of mobile digital fluency, from the three-stage approach cited in Spencer (2015), so that the foundation layer is mobile digital proficiency (also referred to as digital competence), where students learn about mobile environments. This would include learning the basics of how to use the technology for both formal and informal learning tasks and to communicate and collaborate with fellow learners and teachers. This skill level forms the basis for the progression to mobile digital literacy, where students can learn within digital environments, reaching a meta cognitive level where they can independently apply mobile tools to learning. The final level is mobile digital fluency (also referred to a digital intelligence), where the learner is able to fully perceive and leverage all the affordances of mobile learning. Such a layered model recognizes Scarantino's (2003) views of the probability of affordances being perceived. At each level, learners will become aware of more potential affordances of mobile learning. There are transition

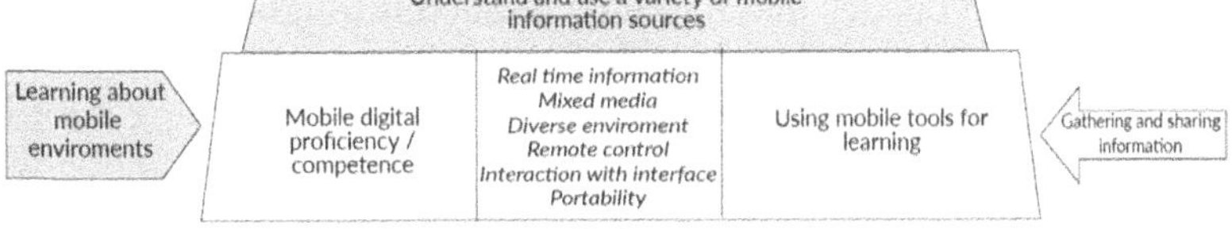

Figure 3.1. The Ziggurat of mobile learning: A proposed mobile learning curriculum structure.

points between these layers, so that learners progress from proficiency to competence when they can understand and use a variety of mobile information sources. Similarly, learners progress to digital fluency when they can operate as mobile digital citizens.

In terms of skills and affordances, students at the digital proficiency level would use mobile tools for learning, utilizing affordances such as accessing real time information, using mixed media in diverse situations, interacting with the interface and perhaps using their devices as remote controllers. Learning activities would center around gathering and sharing information.

At the midlevel of the structure, students will be digitally literate, and able to leverage mobile devices to learn in authentic contexts (UNESCO, 2013). Mobile learners who are digitally literate will be able to identify and gather real-time information, anytime, anyplace (Lai et al., 2007). They will be able to learn with immediacy through various mobile communication channels, engaging and interacting with others directly or as a remote audience.

At the highest level of the structure, students will be digitally fluent with mobile devices and using them to create digital artefacts and exploring mobile entrepreneurship. These artefacts may be outcomes of inquiry (Looie et al., 2014) or mobile applications (CSTA Standards Task Force, 2011). Learning activities will be carried out with empathetic digital citizenship and digital emotional intelligence (Park, 2016; Spencer, 2015). Learning will be contextual and may include multiple indoor and outdoor environments, gathering and using evidence and data.

SUMMARY AND CONCLUSIONS

In this chapter, we have explored the relationship between mobile learning and the curriculum. In doing so we have investigated some representative curricula from several countries, and looked at how a mobile learning curriculum might be constructed. We have noted that, while many curricula mention the use of mobile devices in teaching and learning, these references are superficial and inconsistent. We further note that previous work on mobile learning curricula does not take account of broader digital fluency that might be applied throughout the curriculum. To present a new model of the relationship between mobile learning and the curriculum, we have integrated a set of mobile learning affordances into a layered model that is built upon a journey towards digital fluency, associated with certain skills and learning activities. The purpose of this model is to inform curriculum design such that appropriate elements of mobile learning may be applied into curricula at different levels of learning development, so that mobile digital proficiency, literacy and fluency can be developed over time and in different parts of a learning journey. The proposed model does not focus on specific topics or themes. Rather, it attempts to identify the most important features of mobile learning that might be clearly addressed by a range of specific learning activities, the tasks of which can be mapped to higher level skills. It is hoped that this model allows for mobile learning to be seen as an integral component of contemporary digital teaching and learning rather than as a siloed peripheral subject area. It is anticipated that future research will link this conceptual framework to practice, with a focus on supporting education leaders to better integrate mobile learning within their schools and other learning institutions.

REFERENCES

Aesaert, K., Vanderlinde, R, Tondeur, J., & van Braak, J. (2013). The content of educational technology curricula: a cross-curricular state of the art. *Education Technology Research and Development, 61*,131–151.

ACARA. (2013). *Draft Australian Curriculum: Technologies*. Retrieved from http://consultation.australiancurriculum.edu.au/Static/docs/Technologies/Draft%20Australian%20Curriculum%20Technologies%20-%20February%202013.pdf

ACM. (2004). Computer Engineering 2004: Curriculum guidelines for undergraduate degree programs in computer engineering. Retrieved from http://www.acm.org/education/education/curric_vols/CE-Final-Report.pdf

ACM. (2008). Information Technology 2008: Curriculum guidelines for undergraduate degree programs in information technology. Retrieved from http://www.acm.org//education/curricula/IT2008%20Curriculum.pdf

ACM. (2010). IS 2010: Curriculum guidelines for undergraduate degree programs in information systems. Retrieved from http://www.acm.org/education/curricula/IS%202010%20ACM%20final.pdf

ACM. (2013). computer science curricula 2013: curriculum guidelines for undergraduate degree programs in computer science. Retrieved from http://www.acm.org/education/CS2013-final-report.pdf

ACM. (2014). Software Engineering 2014: Curriculum guidelines for undergraduate degree programs in software engineering. Retrieved from http://www.acm.org/binaries/content/assets/education/se2014.pdf

Ananiadou, K. & Claro, M. (2009). 21st century skills and competences for new millennium learners in OECD countries (Education Working Papers #41). Peru: The Organisation for Economic Co-operation and Development.

Android Developers. (n.d.). *Sensors Overview.* Retrieved from https://developer.android.com/guide/topics/sensors/sensors_overview.html

Bawden, D. (2008). Origins and concepts of digital literacy. In C. Lankshear & M. Knobel (Eds.), *Digital literacies: Concepts, policies and practices* (pp. 17–32). New York, NY: Peter Lang.

Botha, A., Batchelor, J., Traxler, J., de Waard, I., & Herselman, M. (2012). Towards a Mobile Learning Curriculum Framework. In Proceedings *IST-Africa 2012 Conference.*

Bower, M. (2008). Affordance analysis: Matching learning tasks with learning technologies. *Educational Media International, 45*(1), 3–15.

Brown, J. (2015). Complexities of digital technology use and the teaching and learning of function. *Computers & Education, 87*, 112–122.

Buchanan, R. (2011). Paradox, promise and public pedagogy: Implications of the federal government's Digital Education Revolution. *Australian Journal of Teacher Education (Online), 36*(2), 67.

Cardone, G., Cirri, A., Corradi, A., Foschini, L., & Maio, D. (2013). MSF: An Efficient Mobile Phone Sensing Framework. *International Journal of Distributed Sensor Networks*. Article ID 538937.

Ciampa, K. (2014). Learning in a mobile age: an investigation of student motivation, *Journal of Computer Assisted Learning, 30*(1), 82–96.

Chen, B., & Denoyelles, A. (2013). Exploring students' mobile learning practices in higher education. *Educause Review.* Retrieved from http://www.educause.edu/ero/article/exploring-students-mobile-learning-practices-higher-education

Conole, G., & Dyke, M. (2004). What are the affordances of information and communication technologies? *ALT-J, Research in Learning Technology, 12*(2), 113–124.

CSTA Standards Task Force. (2011). *K–12 Computer Science Standards*. Retrieved from http://csta.acm.org/Curriculum/sub/CurrFiles/CSTA_K-12_CSS.pdf

Department for Education. (2013). *Computing programmes of study: Key stages 1 and 2 national curriculum in England.* Crown copyright.

Digital Education Advisory Group. (2013). *Beyond the Classroom: A New Digital Education for Young Australians in the 21st Century.* Retrieved from https://docs.education.gov.au/system/files/doc/other/deag_final_report.pdf

Eaton, D. (2015). The tech education challenge in New Zealand. *Hewlett Packard Enterprise*. Retrieved from http://www.nztech.org.nz/wp-content/uploads/2015/10/16338-HPE-Education-Point-of-View-Document-Oct-2015_4.pdf

Gates Foundation. (2011). *Gates Foundation announces portfolio of innovative grants to develop new teaching and learning tools that support teachers and help students.* Retrieved from http://www.gatesfoundation.org/Media-Center/Press-Releases/2011/04/Gates-Foundation-Announces-Portfolio-of-Innovative-Grants-to-Develop-New-Teaching-and-Learning-Tools-that-Support-Teachers-and-Help-Students

Havelka, S. (2013). Mobile information literacy: Supporting students' research and information needs in a mobile world. *Internet Reference Services Quarterly, 18*(3–4), 189–209.

ITL Research. (2012). *21st Century Learning Design Rubrics*. Retrieved from https://education.microsoft.com/GetTrained/ITL-Research

Parry, D. (2011). Mobile perspectives: On teaching mobile literacy. *Educause Review, 46*(2), 14.

Lai, C.-H., Yang, J.-C., Chen, F.-C., Ho, C.-W., & Chan, T.-W. (2007). Affordances of mobile technologies for experiential learning: the interplay of technology and pedagogical practices. *Journal of Computer Assisted Learning, 23*(4), 326–337.

Leask, M., & Meadows, J. (Eds.) (2000). *Teaching and learning with ICT in the primary school*. London, England: Routledge Falmer.

Looi, C.-K., Sun, D., Wu, L., Seow, P., Chia, G., Wong, L.-H., Soloway, E., & Norris, C. (2014). Implementing mobile learning curricula in a grade level: Empirical study of learning effectiveness at scale. *Computers & Education, 77*, 101–115.

Male, T., & Burden, K. (2014). Access denied? Twenty-first-century technology in schools. *Technology, Pedagogy and Education, 23*(4), 423–437.

Malone, T. W., & Lepper, M. R. (1987). Making learning fun: A taxonomy of intrinsic motivations for learning. In R.E.Snow & M. J. Farr (Eds.), *Aptitude, learning, and instruction: III. Conative and affective process analyses* (pp. 223–253). Hillsdale, NJ: Erlbaum.

Ministry of Education. (2017). Digital Technologies Hangarau Matihiko. Retrieved from https://education.govt.nz/assets/Documents/Ministry/consultations/DT-consultation/DTCP1701-Digital-Technologies-Hangarau-Matihiko-ENG.pdf

Naismith, L., Lonsdale, P., Vavoula, G., & Sharples, M. (2004). Literature review in mobile technologies and learning. *Futurelab Report 11*. Retrieved from https://lra.le.ac.uk/bitstream/2381/8132/4/%5b08%5dMobile_Review%5b1%5d.pdf

Ng, W. (2013). Conceptualising mLearning literacy. *International Journal of Mobile and Blended Learning, 5*(1), 1–20.

NZ MoE. (n.d.). Towards digital fluency. *New Zealand Ministry of Education*. Retrieved from http://www.education.govt.nz/assets/Uploads/Towards-Digital-Fluency.pdf

OLPC. (2016). *One laptop per child*. Retrieved from http://one.laptop.org/

Pachler, N., Cook, J., & Bachmair, B. (2010). Appropriation of mobile cultural resources for learning. *International Journal of Mobile and Blended Learning, 2*(1), 1–21.

Park, Y. (2016). 8 digital life skills all children need—and a plan for teaching them. *World Economic Forum*. Retrieved from https://www.weforum.org/agenda/2016/09/8-digital-life-skills-all-children-need-and-a-plan-for-teaching-them

Parsons, D., Thomas, H., & Wishart, J. (2016) Exploring Mobile Affordances in the Digital Classroom. In Proceedings *IADIS International Conference on Mobile Learning 2016*.

Pegrum, M., Oakley, G., & Faulkner, R. (2013). Schools going mobile: A study of the adoption of mobile handheld technologies in Western Australian independent schools. *Australasian Journal of Educational Technology, 29*(1), 66–81. Retrieved from http://citeseerx.ist.psu.edu/viewdoc/download?doi=10.1.1.684.3653&rep=rep1&type=pdf

Rasinen, A. (2003). An analysis of the technology education curriculum of six countries. *Journal of Technology Education, 15*(1). Retrieved from http://scholar.lib.vt.edu/ejournals/JTE/v15n1/rasinen.html

Scarantino, A. (2003). Affordances explained. *Philosophy of Science, 70*(5), 949–961.

Spencer, K. (2015). What is digital fluency? *CORE Education*. Retrieved from http://blog.core-ed.org/blog/2015/10/what-is-digital-fluency.html

Starkey, L. (2010). Teachers' pedagogical reasoning and action in the digital age. *Teachers and Teaching, 16*(2) 233–244

UNESCO. (2013). *UNESCO Policy Guidelines for Mobile Learning*. Retrieved from http://unesdoc.unesco.org/images/0021/002196/219641e.pdf

Victoria State Government. (n.d.). *Victorian curriculum: Digital technologies*. Retrieved from http://victoriancurriculum.vcaa.vic.edu.au/technologies/digital-technologies/introduction/rationale-and-aims

CHAPTER 4

U.S. URBAN TWEENS TALK ABOUT MOBILE PHONES

Dominant Discourses, Danger, and the Importance of Information Seeking

Sousan Arafeh, Michael Kuszpa, Meghan Weller, and Thomas Mitchell
Southern Connecticut State University

INTRODUCTION

Networked and mobile technologies have become enduring features of the fabric of daily life and there is a wide-ranging body of research exploring their various roles and impacts. It ranges in focus from discussions of mobile technologies themselves; to descriptions and analyses of various software, applications, interfaces, and apps; to explorations of the complex cultural meanings and uses people ascribe to them. The rise and impact of mobile phones—particularly smartphones—has been of great interest. In part, this is because of their increasing ubiquity. In 2016, mobile phone penetration was roughly 80% among the entire 13+ U.S. population, 54% worldwide, and 84% in advanced economies (Poushter,

Mobile Learning: Perspectives on Practice and Policy
pp. 65–92

2016). Mobile phones have been impactful because they afford immediate, personalized, and always-available voice and high-function Internet and app connectivity. Questions have inevitably emerged about how mobile phones are affecting youth and their ability to productively engage in society.

The questions are warranted. Large numbers of adolescents and preadolescents own or have access to mobile phones. As a result, they also have unprecedented access to the many intended and unintended impacts that these mobile phones make possible. There has been much academic and market research on teen and preteen mobile phone penetration and use worldwide (e.g., GSMA and NTT Docomo, 2012; Ling & Haddon, 2008; Poushter, 2016). As this research has progressed, we have found that certain narratives about youth and phones have become dominant (e.g., youth use phones primarily for social media and gaming, they engage in deep positive and negative relational activity on phones, phones are a source of tension between parents and youth due to safety concerns, there is a mobile phone access and use "digital divide") (Common Sense Research, 2016a & 2016b; Lenhart, 2009, 2015). These narratives do describe key features of the tween and teen mobile phone experience. However, based on findings from our inquiry, we suggest that certain aspects of these narratives may characterize youth's mobile phone perceptions and experiences too glibly, monolithically, or without attention to certain very complex details that emerge from each and every user's particular story. We believe that these dominant narratives only scratch the surface of how youth are using mobile phones for learning, particularly for self-directed information-seeking relevant to their daily lives. In addition, we warrant that implicit researcher bias, or not-so-implicit bias, plays a part in the ways that data are interpreted and, thus, how research narratives are formed and mobilized.

In this chapter, we seek to disrupt such dominant mobile phone and mobile learning narratives. We do this by foregrounding the certain specific mobile phone learning insights of "urban" tweens who participated in the exploratory qualitative study we conducted in the Summer and Fall of 2016. This population is statistically at higher risk than other youth along a wide range of indicators and is also underrepresented in the research on mobile phone use. Yet, from our personal experiences as urban middle school teachers, we know that these youth are learning, growing, and striving—often very successfully—despite having to deal with structural and institutional racism, stereotyping, cultural misunderstanding, fraught personal and interpersonal development, and other types of environmental and social/emotional flak. In such a context, and with what is currently known about the importance of mobile phones in urban youths' lives, further research detailing how they are using these devices for successfully learning and living is certainly warranted.

We chose to use grounded theory in the conduct of this study, which made it possible for us to bracket any type of formal theoretical framework that might predispose a particular interpretation (Glaser & Strauss, 2009; Strauss & Corbin, 1998). Grounded theory's inductive method allowed us to interpret what our tweens shared in new ways, led us to interrogate multiple literatures typically not in conversation with each other, and come up with findings that challenge dominant narratives and discourses regarding urban youth, technology, and mobile phones. Through our analysis, we have identified and advanced new theorizations of youth mobile phone use for learning; specifically around the construct of "information seeking."

Our interest in eliciting and spotlighting the mobile phone experiences of this age group of adolescents living in large U.S. cities was motivated by the desire to address three specific issues. First, we wanted to learn more about the perceptions and experiences of developmentally distinct youth in the middle grades. Most education, census, and technology research combines this age group within youth ages 9 through 18 so that the particular experiences of middle grades students are unclear (e.g., U.S. Census Bureau, U.S. Department of Education, Pew Research Centers, Common Sense Media). A second motivation was that we wanted to learn more about the perceptions and experiences of urban youth. Considering how much this demographic is invoked in education and communications discourses, we found very little U.S. research focused on, or emerging from, urban tweens and teens specifically. Lastly, we wanted to see if any of the perceptions and experiences of urban youth elicited in our focus groups would disrupt or challenge dominant discourses regarding youth and digital technology for learning; specifically, for learning with mobile phones.

In what follows, we share what we have learned with the hope that it is relevant for technology and education sector industry members; first-line adults such as parents, teachers, parents, and community support workers; researchers in multiple disciplines; as well as education and technology policymakers.

THE CASE AND CONTEXT FOR STUDYING THE MOBILE LEARNING OF URBAN TWEENS

What Is Mobile Learning?

Understanding mobile phone access, use, and learning for the tween demographic is important for a number of reasons. Virtually all forms of interpersonal communication and engagement with health, employment, education, recreation and other key social institutions use mobile/digital technology. Formal educational institutions use digitized texts supplemented with online materials; learning management systems organize and track curricula and homework progress; laptops and tablets (often coupled with interactive whiteboard capacity) allow teachers to personalize students' learning and teach in locales outside of the classroom or school building; and ever more robust data processors and video and memory capacity offer access to information in a variety of text and multimedia formats. The curricula, materials, learning tools, and pedagogy have all been influenced (Greenhow & Lewin, 2016). The informal learning that takes place through games, social media, individual exploration, or person-to-person communication has also been affected (Pew Research Center, 2016b; Rogoff, Callanan, Gutiérrez, & Erickson, 2016). In this hyperconnected media ecology, access to, and use of, "smart" mobile cell phones, tablets, ultraportable laptops, and other WiFi and/or 3G/4G connectable information, communication, and entertainment devices end up being key educational opportunity and equity/social justice issues. For youth who cannot engage deeply with these technologies, the digital divides are real and affect their social and career success.

"Tweens" in the Middle Grades

School-aged youth sit squarely in the middle of this context where access to networked technology is crucial for development and success in the 21st century. Critically understanding their engagement is key for ensuring they have what they need to succeed. As mobile phones—particularly smartphones—become a primary way to access and share school, career, and personal information, it is important to understand how youth use these phones to learn by connecting to the Internet, other people, and fundamental social institutions.

We focus on youth in the middle grades ages 12–14 because the classical literature[1] and more recent scholarship agree that the middle school years are developmentally unique in the human lifespan (Mertens & Anfara, 2007). In grades five through eight, youth face significant growth that shapes the meaning and purpose they accord to life as they begin the fraught transformational journey from childhood to adulthood. It is during this time that they experience unprecedented cognitive, metacognitive, social, emotional, moral, psychological, spiritual, and physical growth. They also experience significant changes in their identity development, self-concept, and relationships (Wigfield, Lutz, & Wagner, 2005). Youth, likewise, increase risky behaviors and begin a decline in their prosocial attitudes and behaviors during these years. A solid mix of independence and support is needed to help them successfully make the transitions required by this difficult, but formative, time of life. This is particularly so for students from at-risk backgrounds, since the opportunity and achievement gaps increase and solidify in middle school (Kennedy, Brinegar, Hurd, & Harrison, 2016).

Research on Mobile Phones and Youth

There is no doubt that mobile phones have become increasingly influential and entrenched in our lives. Ninety-five percent of Americans own a cell phone of some kind (Poushter, 2016a). In 2015, 13% of all U.S. households with an income of $30,000 or less were "smartphone-only" (e.g., households that use their smartphone for all the connectivity-related activities they used to do on satellite or hardwired Internet, television, or telephone lines) (Horrigan & M. Duggan, 2015). This is nearly twice the 8% smartphone-only connectivity cited in 2013. According to this same Pew report, "Those who have smartphones only or have limited online access options tend to be younger, lower-income, and are more likely to be non-White. They also are more likely than other users to run up against data-cap limits that often accompany smartphone service plans, and more frequently have to cancel or suspend service due to financial constraints" (p. 2).

Youth smartphone use varies based on demographic characteristics with top uses being social media, music, games, online video, and web browsing. As of 2015, 24% of tweens ages 8 to 12 and 67% of teens ages 13 to 18 had their own smartphone (Common Sense Research, 2015). We can extrapolate from this data to estimate that 6% of 8 year olds, 22% of 11 year olds, 29% of 12 year olds, 36% of 13 year olds and 43% of 14 year olds—nearly half—might own a mobile phone if data were not aggregated. Ninety-one percent of teens ages 13 to 17 reported accessing the Internet on some type of mobile device in 2015 (Pew Research Centers, 2015). When broken down by income, 51% of lower income tweens/teens, 69% of middle income tweens/teens and 78% of higher income tweens/teens in the study had smartphones—a *27% difference* in smartphone ownership between lower income and higher income tweens/teens. 22% of all tweens and 58% of all teens reported that they used a smartphone for the full range of digital activities in which they engaged (Common Sense Research, 2015). This study did not report on phone ownership by race.

Urban youth are at risk of being information impoverished due to limits to their voice calling, texting, and Internet access as well as higher costs of use (Brown, Campbell, & Ling, 2011). Lacking access to these communication avenues can severely impact youths' educational and personal growth, a situation that feeds into the well-documented "digital divides" that exist based on disparities in income, geography, technology type, service level, and cultural factors. There is a technical access divide (Council of Economic Advisors, 2015), but also a "second-level divide" based on skill disparities (Hargittai, 2001; Park, 2014), and a "participation divide" whereby less skill in using connected technologies limits access to online information and services (Hargittai & Jennrich, 2016) and results in "information poverty." Experiencing one or all of these divides result in longer term negative social and economic consequences.

What is qualitatively known about youth and cell phones stems largely from empirical studies on social and cultural aspects of U.S. tween and teen use of technology/new media. This research considers uses for gaming, listening and creating together, socializing on social media, and "connected learning," but does not focus on urban youth specifically (e.g., boyd, 2014; Ito et al., 2007/2010). Ling and Haddon's (2008) comprehensive review of studies across the globe on child-parent and peer understandings and uses of mobile phones also does not focus specifically on urban youth. It does, however, tease out key issues relating to youth and mobile phones such as identity construction, development of independence, privacy and safety, and parent concern about immediate two-way information and communication.

According to the authors, this body of research on youth mobile phone practices shapes, and is shaped by, the social construction of childhood via different discursive or cultural approaches to what constitutes appropriate child/youth roles, behaviors, legitimate actions, and legitimate locales for action. Part of these behaviors are those where youth exceed what is socially accepted for their age and station. This is a point also made by Ling's (2005) research on Norwegian teens' efforts to emancipate themselves from the grasp of adult expectations by developing identities and relationships with peers, or even experiment with crime, drugs, and other difficult or dangerous activities. While sometimes poorly conceived or scary, these independence-seeking behaviors, and mobile phones, are part of youth's developmental arc.

Research on Mobile Phones and Urban Youth

We were surprised to find very little focused research on mobile phone access and use by U.S. urban, low income youth and their families; especially from the perspective of tweens themselves. Rather, research tends to detail statistics of ownership and broad use patterns and increasingly includes "user profiles" as ways to qualitatively represent user characteristics and practices. From this research, we know that urban teens are more likely to own and use cell phones as their primary portals for Internet access than middle class teens, which is a fact that has important educational and social implications (e.g., Brown et al., 2011; Common Sense Research, 2015; Pew Research Centers, 2015). MEE Productions (2008) found that 92% of urban Los Angeles youth owned cell phones in 2008. Rice, Lee, and Taitt's (2011) public health study of homeless teens in urban Los Angeles found that upwards 62% of homeless youth in the study had a cell phone, but only 40% had phones in working condition. Youth in shelters and older youth were more likely to own or have access to phones. In addition to using the phones for kin-keeping with friends and family, homeless youth also used their phones instrumentally for health and mental health support, employment, education, connecting with parole officers, and other juvenile justice system contacts. They did this despite the difficulty of having and keeping a working, connected cell phone. From a research perspective, cell phones have made it possible to communicate with, intervene, and gather information from homeless youth and families in new ways (Leinen, 2016).

There is limited qualitative information specifically about U.S. urban youth and mobile phones. Such research is either buried within studies' general narrative of teens (boyd, 2014; Ito et al., 2007/2010), analysis of extant primary sources (Lim, 2013; Park, 2014), or located in research outside of communications, education, and digital/new media fields such as health or criminology. (e.g., Gray, Klein, Noyce, Sesselberg, & Cantrill, 2005). The literature holds that mobile phones have special importance for working class youth and their families because often parents are not at home due to work, school, or other necessary obligations. Thus, working class youth (some possibly not urban) are required to take on more responsibility and autonomy in navigating their environment than middle-class youth. Mobile phones are the primary way these parents and youth can keep in touch.

Two large-scale studies provide insight into the media use of diverse, but not necessarily urban, youth and families. *Connection and Control: Case Studies of Media Use Among Lower-Income Minority Youth and Parents* use the tween/teen typology of Heavy Viewers, Light Users, Readers, and Gamers/ Computer Users, Video Gamers, and Mobile Gamers to structure nine cases characterizing what mostly urban 11-15-year-old youth do with digital media (Common Sense Media, 2015). The report *Diverse Families and Media: Using Research to Inspire Design,* also creates learning outcome-focused cases of youth and family media use (e.g., family language learning, connecting to heritage culture) so that media designers can create intergenerational and collaborative conditions for digital media and use (Levinson, Siyahhan, Pressey, & Taylor, 2015). While both of these studies outline diverse families' media experiences and share actual youth and adult voices, neither of them look specifically at mobile phone use. As well, the case characterizations do not fully capture youth phone-based digital media practices and experiences.

In contrast to this U.S. research, research on urban youth and mobile phones in non-U.S. countries is much more robust. In addition to early detailed studies of mobile phones in youth culture in Norway (Ling, 2004), there is research on cell phone use by urban youth in the United Kingdom (Pain, Grundy, Gill, Towner, Sparks, & Hughes, 2005), Botswana (Lesitaokana, 2014), Switzerland (Geser, 2006), Philippines (Portus, 2001), Sweden (Thulin & Vilhelmson, 2009), and Japan, India, Indonesia, Egypt, and Chile (GSMA/NTTDocomo, 2012).

STUDY APPROACH AND METHODS

This exploratory qualitative study investigates urban tweens' mobile device perceptions and experiences as they were shared with us in three focus groups held during the summer of 2016. It also critically considers narratives and discourses of the mobile phone experiences of youth through the review of selected literature. A grounded theoretical approach was utilized so that we could build theory by eliciting themes inductively from the data collected (Glaser & Strauss, 2009; Strauss & Corbin, 1998). Once focus group data collection and analysis took place, we explored research from multiple disciplines relevant to our findings and rubbed the focus group themes against it. This allowed us to refine our understandings of urban youth and mobile phones and to develop new understandings.

We conducted three focus groups in which we elicited tweens' beliefs, actions, and expectations regarding their specific uses of mobile phones for "learning." Our learning-related research questions were:

- What are the most common things you do on your mobile phone?
- How do you use your mobile phone for learning?

We asked these questions broadly in a way that would pertain to all types of mobile phones; however, smartphones emerged as the primary phone type these students engage with for learning.

This study was conducted at three separate middle pre-Kindergarten to eighth grade interdistrict magnet schools in two large urban cities in Southern New England. The cities have a median household income of $37,000 and $41,800 respectively with poverty levels of 26.6% and 22% (ctdata.org). These two cities are more violent than 6% and 19% of all cities in the United States (Neighborhoodcout.com) with dangers being more likely for those in high crime neighborhoods. Table 4.1 provides demographic information about each school.

We recruited and conducted three gender-balanced, racially, and ethnically diverse focus groups comprised of a total of 18 seventh and eighth grade urban tweens. We spoke with 7 seventh graders and 11 eighth graders of whom 11 were males and 7 were females. Their ages ranged from 12 to 14 and their racial/ethnic breakdown is shown in Table 4.2.

Participants were recruited through fliers and word of mouth. They consisted of all students who expressed interest, were able to attend the focus group session at their school, and were able to produce a signed parental/guardian consent form. Thus, participants were self-selected and motivated. Each focus group was roughly one hour long, audiotaped, transcribed, and then subjected to thematic analysis (Boyatzis, 1998) using open and axial coding in a constant comparative method (Glaser & Strauss, 2009; Strauss & Corbin, 1998). Once the focus group data were collected and analyzed, we identified literature for our general literature review of the topic (e.g., youth and mobile phones, urban youth and mobile phones, learning with mobile phones). We also sought out literature pertaining to themes proffered by participating youth (e.g., themes of learning about sensitive or dangerous topics led us to literature on social deviance and information seeking). We identified and selected only scholarly works in the form of books, peer reviewed articles, and government or non-governmental organization reports. These were identified

Table 4.1. Selected Demographics of Participating Schools

	Total School /Middle School Population	*% Black /African American*	*% Hispanic /Latino*	*% White*	*% Asian & American Indian*	*% Free /Reduced Lunch*
School 1	470/90	60%	20%	18%	< 3%	52%
School 2	470/72	31%	42%	21%	<5%	23%
School 3	650/189	41%	48%	7%	<2%	52%

Table 4.2. Demographics of Student Focus Group Participants

Race	*Male*	*Female*	*% of Participants*
African American	4	2	33
Hispanic	5	4	50
White	1	0	5
Multiracial	1	1	11
Totals	11	7	100

using Boolean search strings of relevant terms in the GoogleScholar, JSTOR, and Academic Search Premier search engines. While crafting the literature review, we also conducted thematic analysis to identify dominant and, more importantly, submerged themes in the research warranting further consideration (e.g., the lack of focus on urban youth experience of mobile phones for learning, information seeking absent from dominant narratives and discourses of mobile phones and youth). Themes relating to mobile phone access and learning are considered here along with our own theoretical developments on this topic.

FINDINGS

Findings From the Focus Groups

In what follows, we report selected findings from the three focus groups that relate specifically to mobile-phone-facilitated learning.

Urban Tweens Report That the "Digital Disconnect" Persists: Urban Youth Do Not Use Mobile Phones For Learning at School

Most focus group participants reported that they are not allowed to use their mobile phones for learning in school, except in unique situations. This finding extends the unfortunate trend of schools not allowing students to access technology in school for learning (Levin & Arafeh, 2002). This is odd because 24/7 nearly ubiquitous Internet and interdevice connectivity via broadband, Wi-Fi, and 3G/4G cellular networks are available. In addition, personalized learning and STEM (Science, Technology, Engineering, and Mathematics) learning has been strongly advocated by government, industry, and education for crucial career and human capital/economic reasons. Mobile technology can also enhance the curriculum by providing access to information otherwise inaccessible in schools. Yet, mobile phones are often banned. Here is an excerpt from one school district's 2016 student/parent handbook: "If students bring electronic devices such as cell phones tablets, iPods, mp3 players, and so forth, to school, the devices should be secured in a locker so as not to disrupt the educational process.... Students may be suspended from school if they persist in using electronic devices inappropriately in the school building" (District X 2016 Student/Parent Handbook). Tweens in two of the schools reported needing to comply with this type of policy. They shared that by complying they not only lose out on multimedia non-place-based learning resources afforded by mobile phones, they also risk not being able to fulfill family responsibilities for parents, siblings, or others or who need to keep abreast of what is happening in their neighborhoods for safety and other reasons. This can have devastating results when, for very reasonable reasons, they may violate the no phone policy. If they are caught with their phone repeatedly, the no device policy can quickly result in suspension and snowball into disproportional disciplining of students.

Urban Tweens Displayed Luke-Warm Interest in Social/ Recreational Mobile Phone Behavior and Prioritized Entertainment, Kin-Keeping, and Emergency Use

All of the youth in our focus groups reported owning or having access to mobile phones and engaging in social media and game play of some kind. However, most shared an ambivalence or nonchalance about their phones and social media uses and did not foreground these as mobile-phone-facilitated learning activities. This seemed to suggest that the youth are less connected to their mobile phones in this way than might be expected from the dominant discourse of the current literature. Instead, entertainment and keeping connected with family and friends emerged as important for nearly all participants. Interestingly, "kin keeping" has always been a core reason for using communication technologies like phones and e-mail (Arafeh, 2000). Participating tweens stated they use multimedia content from the phones as both fodder for connection as well as reminiscence of such connection.

Student 2M: I'm personally not attached to my phone. I could throw my phone away and I wouldn't feel anything. [...][2] My phone is just something that I just use just to call and to be entertained.

Student 5M: I'm not really so attached to my phone, but the only thing that matters in my phone is the pictures because it reminds me of memories and all.

Student 4F: I mostly use my phone because I feel like if I don't use it and text people I'm going to miss out on some stuff and get left out [like] a new dance comes out [or] new videos and people talk about it and I'm late. (Focus Group #1)

Urban Tweens Reported Instrumental Information Seeking as the Primary Component of Their Mobile-Phone-Facilitated Learning

Our urban tweens spoke about seeking information "all the time" using their mobile phones. They regularly use their mobile phones to support school-related concepts or processes when not in school. They also use their phones for personal learning on topics of interest, and for communicating with others in English and other languages. What is not evident in the in the current literature is how these urban youth present as highly confident and competent information seekers. These rather young youth situationally identify and access information of all types through the reliable, typically unlimited, and immediate access their mobile phones offer via audio, video, and the ability to call or text friends, family, or other resource people. In fact, it is interesting that, in the brief course of the two dialogues below, our 12- to 14-year-old tweens talked of accessing books, news, and videos for information and examples of content or processes. They reported using Google, Netflix, the British Broadcasting Company (BBC), the NBA app, dictionary apps and websites, GoogleTranslate, and YouTube. They describe these searches as highly interest-driven and independently-initiated for both school and personal reasons. Likewise notable is the degree to which they report using their phones to access both English and non-English vocabulary and phrases for school and for immediate interpersonal communication in school, home, and other venues.

Student 2M: I use [my mobile phone] for calling, texting, and reading. Like books ... if I can't get them from the library, I just try to get them for my phone.

Student 6M: I use my phone for everything. For research. I use my phone for texting my friends, calling for emergencies, watching YouTube, Netflix, yes.

Student 6M: I have BBC [British Broadcasting Company] on my phone like the app, but it's sort of like halfway since you need WiFi for it for it to update. Usually I have BBC on my phone. It sends an alert if anything happens. I use it basically for like projects or anything that comes up in school ... I'll look it up and see what it's all about.

Student 2M: Google Chrome. I use it for research. And I have an app on my phone. It's like a dictionary app, but it lets me know if there's a new word that exists so I can learn it and its definition. [...]

Student 10M: I use my phone a lot for learning. When I'm stuck. I'm not really used to math so I'll always look it up in Google and that really helps me a lot. It shows me examples and all. (Focus Group #1)

Student 1M: I like learning about history, so I look up history documentaries about world wars and other things like that.

Student 2M: I was using Google Slides and I downloaded it on my iPad because I had this social studies project to do. I finished it on my iPad. I also look up stuff for school, like reading, math, and science.

Student 3M: Sometimes I search up NBA things, highlights and stuff like that to help me in the way like that, or Google Docs to help me with my homework and stuff so I can finish my projects and stuff like that.

Student 4M: I go on YouTube sometimes to watch *Most Dangerous Prisons* because I like watching that kind of stuff. Not in a criminal way, but I like seeing what happens in jail and stuff. In school, we're reading about *Monster* [Meyers, 1999] and this kid, he's in jail. We're reading that book. I started watching prison stuff and how they be doing them, like what food they be getting, like dog food and stuff.

Student 5F: I have the Netflix app, and in there it has *The World's Deadliest*. It shows about animals and how they attack other animals, or different strategies that they use. I think it's pretty cool. We're so used to being around people. I don't really go to [the] actual [places] where you see animals and how they react. [I have seen animals] only in zoos, but that's different when it's actually in the wildlife. I think it's interesting to see how they do certain things and use different strategies to get what they want.

Student 6F: I use Safari for [English] vocabulary because I'm really bad at that. I first ask my older siblings and then if they don't know then I have to search it. [...]

Student 3F: The only reason why I use Google Translate is because my parents say so many Spanish words that I don't understand, so I search it up. [...]

Student 4M: I use it for [Spanish] words and phrases. The gym I go to ... there's a lot of Spanish speakers. If I want to give them a compliment I say, "You are strong," or "You hit hard." I look it up on Google Translate. (Focus Group #3)

Urban Tweens Seemed to Prefer Video, Not Text, as Their Mobile Phone Information Source for Learning About Content and Processes

One of the more surprising findings was the degree to which focus group participants cited YouTube, Netflix, or other streaming video as a key source for information of all kinds. They discussed using video of different types for learning content, but also for learning technique and strategy (e.g., news; movies and history documentaries; video on prison life, animal behavior, song lyrics and acting technique, and models of various physical workouts). Focus group participants valued both the instrumental information these provided as well as the visual modeling of how such information could be applied. In the case of schoolwork, participants held that this was not cheating. Rather, they felt video tutorials were key learning tools utilized by family, friends, teachers, and themselves at varying times. Students who talked about using video to explain and model academic content seemed proud of their ability to seek guidance toward mastery and not just look for the answers, which shows a level of independence and maturity perhaps not expected of youth of this age. As well, students valued their independence and autonomy in being able to really express and experience their personal interests through accessing relevant information via their mobile phones.

Student 2M: Last year, I remember a lot of the time if I didn't understand something, my dad would help me by looking it up and showing me the video of how to do it, or he would look at the video and then try and explain it better to me. (Focus Group #1)

Student 3M: I'm not being [a cheater. I use videos] to answer my questions. I'm trying to see how to do it in a way. Sometimes I'll be confused on how to do it in a way, so I'll be watching videos on YouTube on how to do it, but I don't answer the question fully. (Focus Group #3)

Student 1F: If I really can't solve something, if I need help, I will look up a YouTube tutorial to see how to do that kind of problem and then I used to do it for IXL and stuff. And maybe I could read books on my phone if I had it for homework.... I forgot to say that I love acting and singing, so that's why I watch movies and stuff all the time. I also use my phone to listen to music and for lyric videos to sing the songs ... and if I see a show and I like how someone plays the character or I just like it, I'll search up who plays it.

Student 2M: When I want to work out, to learn how to work out, I watch The Rock [on YouTube]. He's a role model. When I go to watch him, I'm outside with my brother to like do some of the workouts like he does. (Focus Group #2)

Certain Male Tweens Reported Using Their Mobile Phones to Seek Information About Troubling or Dangerous Matters (i.e., Guns, Prison, Dark Web)

We were not surprised that some of the male participants reported accessing content largely "inappropriate" for middle school-aged children. However, when, in three different focus groups, three male youth disclosed that they used mobile phones to access information about guns, prison, and the dark web we found ourselves starting to weave a narrative that was potentially stereotypic or contained negative stereotypic, but also possibly true, elements. Were they using the dark web to hide illicit activity? Were they trying to learn about guns so as to buy one? Do they live in rough neighborhoods or are they involved in gang or other nefarious street behavior? Should this information be reported to the school and/or their adult caregivers? Are these youth safe?

After checking in to be sure that their revelations communicated no immediate danger, we listened closely and felt their interest was actually grounded in an earnest desire to learn about the dangers or issues related to guns and prisons so as to be able to engage with them in an informed manner. We arrived at this conclusion, in part, because these students shared this information in a public setting in response to the question of how they use their mobile phones for learning—it was not something they were hiding. Our further analysis of their statements lead us to believe that the cases reported here constituted learning and, because of the identity location of these boys (e.g., urban, lower income, minority), *may* have also driven interest. We strove to be highly cognizant that our interpretations could have negative stereotypic biases.

The interest in learning about prison life exemplified in the quote above, was sparked by a school reading assignment. In the case of interest in guns, the impetus was not evident.

> Robert: I'm [Robert] and I'm 13 and I'm in 8th grade and I'm funny and I make friends very good. When I'm on my phone, I'm on YouTube and watching this dude on Demolition Ranch because he does stuff with guns and I like that stuff. Then sometimes I go on deserteagle.com. (Focus Group #2)

In the case of learning about prison, we were not able to follow up to dig deeper into this youth's interest, so our speculations regarding why this youth was interested in prison life were suspect for their assumptions and possible stereotyping. As a way to learn more about this teen's interest, we viewed a number of *Most Dangerous Prison* clips and found that they offered a great deal of information about prison life that we were not aware of. We likewise searched on deaserteagle.com and glock.com and watched multiple episodes of Demolition Ranch. We saw how these websites and YouTube episodes provided interesting empirical information about guns and ammunition while, at the same time, wondering, really, why this tween was interested. In this case, we were able to ask the student why he was so interested in guns and, as it turned out, his interest in gun types, specifications, and performance was driven by career interest.

Facilitator: You were saying that you were looking at gun sites. I'm wondering what's your interest in that?
Robert: I want to be in the FBI. That's why I watch [the TV show] *Criminal Minds*. Most of it is, like, connected.
Facilitator: I see. And you said it's called Desert Eagle? What is that?
Robert: It's one of the big caliber pistols. It's big. Also, I look at glock.com. That's another pistol and sometimes I just watch the Youtube, Demolition Ranch.
Facilitator: Okay, and what's that?
Robert: It's a dude that shoots stuff out of a shotgun or he just shoots stuff with guns.
Facilitator: Is he testing them and seeing what's the best thing? [...]
Robert: [Nods....]
Facilitator: And why does looking at the guns help you with the FBI idea?
Robert: Most of the stuff and FBI mysteries and stuff ... most of it's crime, so you've got to.... Like, if there's a bullet, you have to connect the bullet with the gun and match it up.

While only a snapshot into some of the types of learning urban youth reported through their mobile phones, our analysis of these particular instances did unearth some new themes worthy of consideration by the field. We found that urban tween mobile phone behavior does not necessarily fit the stereotypes and typologies evident in the literature. This is particularly true regarding information seeking. Information seeking emerged as key component of urban tweens mobile-phone-facilitated learning. However, as will be discussed the literature analysis below, information seeking is largely absent from the ICT and digital media use literature. We also found that tweens reported less urgency in their typical—largely social/recreational—mobile phone behavior. Lastly, urban tweens seemed to prefer video as the medium through which to seek both school-related concepts or processes and personally relevant information via their mobile phones. When information about troubling or dangerous matters was sought (i.e., guns, prison), it was done so, ostensibly, for valid school, personal, and/or career purposes.

Findings From the Literature Analysis

Having identified these specific ways our urban focus group participants use their mobile phones for learning, we then looked to the literature to see whether and how it would support what we seemed to be seeing. As is evident from the contextualizing portion of this chapter, we first considered research on mobile learning, tweens in the middle grades, mobile phones and youth, and mobile phones and urban youth. We found that there are certain dominant discourses regarding youth and digital technology, of which tweens and mobile phones are a part, that seem to be limiting possible analyses and interpretations.

Through our analysis of the literature, we identified five dominant discourses. In what follows, we point out the notable absence of a discourse of mobile phones and information seeking even though our study has shown information seeking to be a consistent mobile phone practice. We, thus, conclude information seeking should be included in theorizations of youth mobile phone use. Furthermore, we suggest that certain types of information seeking should be included in the research discourse about urban youth because such practices may be more applicable to the way these youth use mobile phones for learning.

The Structuring Impact of Dominant Discourses of Youth and Digital Technology

As we considered the topic, we felt that what is known about urban youth, mobile phones, and learning must be seen in the context of research on youth and connected electronic/digital networks and devices more generally. Our review and analysis of the literature leads us to suggest that five specific research discourses have dominated how youth learning with digital technology is understood conceptually and methodologically in the United States. These five discourses are as follows: (1) the discourse of counting and categorization, (2) the discourse of game-based learning, (3) the discourse of technology practices and learning in youth culture, (4) the discourse of technology-based learning for education, and (5) the discourse of technology and teen deviance. These discourses have been very helpful for understanding digital technology in the lives and learning of youth. However, we argue they have also limited alternate or additional understandings of youth's digitally-mediated practices as evidenced by some of our findings. In this section, we outline what we perceive to be key dominant discourses with structuring effects on how youth and digital technology like mobile phones are constructed and mobilized.

The Discourse Of Counting and Categorization

With the popularization of personal computers and the Internet in the 1980s, communication and learning could take new forms. Tracking and exploring digital devices, software, and activities were crucial parts of this dynamic period worldwide to provide technical, social, and economic guidance for educators, communities, industry, families, and government. It continues to be important to know about, thus, it was, and continues to be, important to know about technology penetration and use, so technical categorization studies are regularly pursued. Perhaps the most influential and visible U.S. research related to youth and digital culture—specifically communication and media technologies—has been that undertaken by the Pew Research Centers, Common Sense Media's Research arm, the Joan Ganz Cooney Center, and market research firms such as Neilsen and Gallup (Common Sense Research, 2015, 2016a, 2016b; Lenhart, 2009, 2015; Lenhart, Arafeh, Smith, & Rankin, 2008; Lenhart, Ling, Campbell, & Purcell, 2010; Levin & Arafeh, 2002; J. Horrigan & M. Duggan, 2015; Poushter, 2016). These well-regarded studies survey and/or interview nationally-representative samples of youth and adults. Information is gathered about technology use and the meanings behind use as a way to characterize the technical and cultural shifts afoot. Such studies have provided substantial research on device and subscription numbers and general use. They have also facilitated important analytic concepts like the "digital divide," "digital native," the "gaming teen," "media teen," the "social media teen" (possibly addicted and/or headed toward social ineptitude), and the "privacy and safety concerned parent" (often unaware of children's real learning opportunities and practices). Important information regarding digital technology demographics has been provided through these studies, which have set the stage for how such information is typically conceptualized for reporting (e.g., descriptive statistics and user profiles/categorizations).

The Discourse of Game-Based Learning

Research by Seymore Papert and MIT's Learning Lab and the Games + Learning + Society Group at the University of Wisconsin–Madison has explored constructionist learning through programming and games (Gee, 2003; Holden, Dikkers, Martin, & Litts, 2015; Klopfer, 2008; Squire, 2011). This research has engendered a dominant discourse of game-based or gamified learning (e.g., serious games, geolocated "quests") as particularly beneficial or youth-preferred forms of digital learning. Place-based and augmented reality games in informal settings (e.g., communities, museums) fall into this category, which have been significantly enhanced by the development of mobile technology. A number of mobile learning games from this legacy appear as chapters in this very book (see the Table of Contents). What is important about this discourse is that in focusing so squarely on literacy and learning within game play, or through the gamification of educational content and processes, it has overshadowed, or possibly ignored, other forms of learning taking place.

The Discourse of Technology Practices and Learning In Youth Culture

There has also been research focused specifically on qualitatively eliciting and describing the meanings and culture(s) developed by youth in the course of using digital technology. The MacArthur Foundation has been a key supporter of such research (e.g., Ito et al., 2007/2010) through fellowships, HASTAC (Humanities, Arts, Science, and Technology Alliance and Collaboratory), and the Digital Media and Learning Research Hub at University of California–Irvine. Resulting dominant discourses have focused on the rich, generative uses of digital media by youth for various entertainment and instrumental purposes. Key research has focused on the social and participatory aspects of networked culture and its blurring of formal and informal learning spaces, including as augmented reality (boyd, 2014; Ito et al., 2007; Ito et al., 2013; Jenkins, 2006). This discourse has helped the field value the media knowledge and practices of youth.

Ito et al. (2007) coined the terms "messing around" and, as part of this, "looking around," which they characterized as a youth media participation practice of information seeking. However, their conceptualization of this activity is significantly bounded. They recognized and valued "fortuitous searching," youth trolling on the Internet to seek and find information serendipitously, rather than more instrumental information searching activity such as we found in our study. They further bounded their conception by honing in on (1) *how youth are using search engines* to search for information or (2) how youth seek information on *how to use new media*. In this and work that follows, other types of non-new-media-related information are not considered (Hargittai & Hinnant, 2008).

The Disconnected Discourse of Technology-Based Learning for Education

There has been research on the use of digital technology and the Internet for formal education- or school-related learning. As noted, early and later Pew Research Center studies found that computer and Internet technology have not been used well for in-school or for-school learning (Lenhart et al., 2008; Levin & Arafeh, 2002; Rainie & Hitlin, 2005). Since then, computers and tablets have increasingly been incorporated into formal school learning (e.g., 1:1 initiatives, BYOD, Digitized commercial or "open" educational resources), but educational uses of mobile phones has been slow, despite literature and ideas for educational mobile phone uses (e.g., Twitter, homework, data gathering) (Kolb, 2008; Kuznekoff & Titsworth, 2013). This discourse has not been evident in mobile phone and learning research.

The Discourse of Mobile Phones and Teen Deviance

While not a dominant discourse, there is acknowledgment that youth use technology for deviant behavior. Defined as either positive or negative behavior "outside of the norm," technology-enhanced spaces are a perfect place for youth to explore risky identities, develop relationships with people for nefarious reasons, or engage in risky or violent behavior directed at others or themselves (Annine, 2005; Ito et al., 2008). Mobile phones have provided a space for new types of deviant behavior (e.g., sexting, microcoordinating deals and meetings), some of which are developmentally normal parts of defining oneself as an individual (boyd, 2014; Ling, 2005). However, discussion tends to be limited to more palatable types of deviance, which excludes more dangerous practices like those identified in this study (but which we deemed can legitimately have different, even prosocial, motivations).

Notably Absent: A Discourse on Information Seeking

Notably absent from these dominant youth and technology discourses is a strong, formal discourse of youth and digital media-based information seeking. With the advent of the Internet, information seeking about work, school, and other topics of interest increasingly has taken place online via computers and now phones. However, this discourse's absence is particularly puzzling because information and formal and informal learning have always been crucial points of focus. We argue that the disciplinary, content, and methodological foci of the dominant discourses just identified have made it possible to overlook information seeking; even when, for example, the discipline of Library and Information Science has advanced parallel work on this topic and with a strong focus on teens and mobile phones (Agosto & Hughes-Hassell, 2008; Dresang, 2005).

Information Seeking

We have identified information seeking as a missing discourse in ICT and media use research on youth and mobile phones. This is true regarding mundane school and personal topics of interest, but also regarding troubling and potentially dangerous topics. While deeply exploring the literature on information seeking and youth is beyond the scope of this chapter, we believe that plumbing this literature would strongly support our theory building and contribute to current knowledge about how urban youth use their cell phones, or might use them more productively.

Broadly, information seeking is the practice of searching for information, whether such searching is wholly instrumental or more serendipitous. Shenton and Dixon (2003) argue that young people's information seeking behaviors are developmental and that the information and search needs at one age are necessarily different from those at another. This parallels Ling's (2005) developmental characterization of teen mobile phone use for emancipation. In a study by Agosto and Hughes-Hassell (2005), youth desired both school and social information and preferred getting it from people close to them. For these teens, mobile smartphones were the favored method of information seeking because they afforded direct interpersonal connection and online information searching. As well, the shape and nature of these youths' information seeking depended significantly upon context. Every different situation requires a different response regarding how key aspects of the matter in question should be treated (e.g., people, medium, topic, location of information).

When we look at the information seeking literature from the perspective of mobile learning, there seems to be natural connections that offer explanatory power. Thus, further research is needed. We have identified that scholarship in multiple disciplines recognizes and validates that youth legitimately need to explore topics that seem concerning and/or dangerous as ways to better understand and navigate the world around them.

DISCUSSION AND CONCLUSION

We began this chapter by saying that we wanted to disrupt certain dominant mobile phone and learning narratives—and sought to do this, in part, by exploring three gaps in the research literature: perceptions and experiences of mobile phone use for learning by youth in the middle grades, perceptions and experiences of mobile phone use for learning by urban youth, and middle grade urban youth challenges to the dominant discourses of learning with digital technology.

Based on our analyses, we feel it is safe to say that the research literature on youth and technology, including mobile phones, does not focus on youth in the middle grades as a specific developmental age group. Yet, this is a unique life period that deserves more consideration so that adults of all kinds can better support tweens in their development.

We also found that, despite the longstanding dominant discourses focusing on social media and gaming as primary uses and causes for concern, the tweens we spoke with seemed to have a more relaxed, nonchalant relationship to their phones and these practices. It may have been that we were asking specifically about how they learn with their phones and so they downplayed their social and gaming activities. However, we felt that they were actually displaying the kind of mobile phone "taken for grantedness" that Ling (2012) and others have identified. Their phones and affordances are now so woven into the daily fabric of their lives that they do not find them, or the activities conducted with and through them, particularly notable. And while responses in the focus groups for this research did not result in significant discussions of how or why urban youth might experience their mobile phones differently from others, some tweens did say there was a sense that because they lived in the city, their phone uses were more frequent, social, and concerned with safety of different types than youth in other areas.

We were somewhat surprised regarding the degree to which our tweens seemed to relish the use of their phones for instrumental information seeking. They not only disclosed that they use their phones outside of school "all the time" to seek school-related and personal information, but also reported instrumental information seeking as the primary component of their mobile-phone-facilitated learning. They seemed to "search up" new and clarifying information constantly. The upswing in video, not text, as a preferred format for learning was also instructive. This finding would suggest that, in addition to the blurring of formal and informal education along a "learning lives" continuum, using video formats in schools, such as through the flipped classroom approach, could result in increased student engagement and results.

What struck us deeply, and resulted in what we feel has been a very productive grounded theory analysis, was that in addition to using their mobile phones for emergency and safety reasons, students were using them to seek out controversial or dangerous information. While that act is not surprising in and of itself, considering it and the very limited literature related to such information seeking activity has sparked in us an awareness of the degree to which these legitimate information needs are not being met and are barely on the radar. In addition, we felt it important to ensure that our engagement with these practices were not alarmist or diminishing, as the students we were speaking with certainly do live in urban areas that are statistically more likely to include violence and crime; incidents of which are also are known to occur more often among certain populations (e.g., lower income, minority, male)—the very type of students sharing the information. We were able to probe a bit more deeply in one case, which made it clear to us that further research into these types of information seeking activities and sincerely asking why, and with what hoped impact, tweens engage in such activity is crucial. The mobile phone serves as a gateway into important cultural, cognitive, and social-emotional space. Thus, we suggest more research along these lines be undertaken along with teasing out what may be different needs by boys and girls. This information will result in considerations for supporting youth and adults and the systems in which they work: home, school, community, sport, workforce, justice/law enforcement, higher education, and so forth.

While our findings are pertinent for urban youth, we believe that there is the potential for generalizability to youth from suburban and rural locales. However, this will need to be determined by future research. We believe such generalizability is the case regarding the continuing "digital disconnect" in schools where mobile phones and other digital technologies are not being used to effectively support learning. We also believe generalizability is the case regarding a student drop in interest for social media, but a consistent valuing of mobile phone entertainment, kin-keeping, and emergency use. Information seeking also seems a finding that is generalizable. We hold out with reservation, however, that the information interest in guns and prison of certain male focus group participants could possibly be more germane to urban life and interest in controversial and potentially dangerous topics for personal edification, preparedness, and safety.

In our study, participating tweens openly shared their information seeking activity. There was no indication that they were holding information back. They seemed comfortable talking about the kinds of information they seek via their mobile phones, how they identify and access the information, and why they do so. This allowed us to identify five dominant discourses that we feel have bounded the discursive construction of digital technology and media for learning by youth: (1) the discourse of counting and categorization, (2) the discourse of game-based learning, (3) the discourse of technology practices and learning in youth culture, (4) the discourse of technology-based learning for education, and (5) the discourse of technology and youth deviance. Through this research, we have come to believe that these dominant discourses/narratives scratch the surface of youth mobile phone use for learning. In considering these discourses, we determined that part of why they exist is our own—researchers'—implicit bias in interpretation and, as a result, disciplinary and narrative congealing, boundedness, and solidification.

To extend beyond these dominant discourses and contribute to the field in new ways, information seeking should be a (re)new(ed) focus. In particular, we want to draw attention to the more independent, solitary, instrumental ways of seeking information for personal edification and use. And, like Ling's (2012) idea that our understandings of youth mobile phone use are structured by the construction of childhood, we suggest our understandings are also bounded by whether or not we construct our youth as independently motivated capable learners committed to their development and advancement—particularly urban youth. We hold that the practices revealed to us in this study are highly mature, rational practices that youth engage in as a way to effectively learn about, and cope, with their worlds.

As digital space and devices move the boundaries of what "learning" is and how it takes place, it is crucial to understand what types of information seeking and knowledge/skill practice are possible and beneficial. It seems important to continue to listen to tweens describe their learning landscapes and trajectories while being open to hearing about the many types of information needs that resonate with youth for their particular developmental, and always changing lives and contexts.

ACKNOWLEDGMENTS

A heartfelt thank you to the districts, schools, students, and faculty/staff who made this study possible. We are grateful for their willingness to share experiences and insights. By doing so, they have not only contributed to research about urban tweens and how they use mobile phones for learning, but also to development of new theory that broadens the field's understanding of this important topic.

NOTES

1. For example, Freud's and Erikson's psychosocial stage theories, Piaget's cognitive development stage theory, Kohlberg's moral understanding stage theory, Vygotsky's social constructivism, and Bronfenbrenner's ecological systems theory.
2. [...] is used when dialogue has been removed to focus on a particular finding. In all cases, the removal of dialogue does not alter the meaning and intent of the group's conversation.

REFERENCES

Agosto, D. E., & Hughes-Hassell, S. (2005). People, places, and questions: An investigation of the everyday life information-seeking behaviors of urban young adults. *Library and Information Science Research*, *27*(2), 141–163.

Annine, J, D. (2005). Adolescent's perceptions of deviance when using technology: The approaching post-typographic culture. *Current Issue in Middle Level Education*, *11*(1), 19–24.

Arafeh, S. (2000). "Women", telephones, and subtle solidarity: A counternarrative. In A. DeVaney, S. Gance, & Y. Ma (Eds.), *Technology and resistance: Digital communications and new coalitions around the world* (pp. 113–138). New York, NY: Peter Lang.

boyd, d. (2014). *It's complicated: The social lives of networked teens*. New Haven, CT: Yale University Press.

boyd, d., & Hargittai, E. (2013). Connected and concerned: Variation in parents' online safety concerns. *Policy & Internet*, *5*(3), 245–269.

Boyatzis, R. (1998). *Transforming qualitative information: Thematic analysis and code development*. Thousand Oaks, CA: Sage.

Brown, K., Campbell, S. W., & Ling, R. (2011). Mobile phones bridging the digital divide for teens in the US? Retrieved from http://www.mdpi.com/1999-5903/3/2/144

Campbell, S. W., & Park, Y. J. (2008). Social implications of mobile telephony: The rise of personal communication society. *Sociology Compass: The Journal Of The British Sociological Association*, *2*(2), 371–387.

Centers for Disease Control and Prevention. (2009). *School connectedness: Strategies for increasing protective factors among youth*. Atlanta, GA: U.S. Department of Health and Human Services.

Common Sense Research. (2015). *Fact sheet: Teens and smartphones*. San Francisco, CA: Retrieved from https://www.commonsensemedia.org/sites/default/files/uploads/pdfs/census_factsheet_teensandsmartphones.pdf

Common Sense Research. (2016a). *Connection and control: Case studies of media use among lower-income minority youth and parents*. San Francisco, CA: Common Sense Media. Retrieved from https://www.commonsensemedia.org/sites/default/files/uploads/research/csm_connection_and_control_research2016_1.pdf

Common Sense Research. (2016b). *Technology addiction: Concern, controversy, and finding balance*. San Francisco, CA: Common Sense Media. Retrieved from https://www.commonsensemedia.org/research/technology-addiction-concern-controversy-and-finding-balance

Council of Economic Advisors (2015). *Mapping the digital divide*. Retrieved from https://obamawhitehouse.archives.gov/sites/default/files/wh_digital_divide_issue_brief.pdf

Dresang, E. T. (2005). Access: The information-seeking behavior of youth in the digital environment. *Library Trends*, *54*(2), 178–196.

Gee, J. P. (2014). *What video games have to teach us about learning and literacy*. New York, NY: Palgrave Macmillan.

Geser, H. (2006, April). Pre-teen cell phone adoption: Consequences for later patterns of phone usage and involvement. *Online Publications: Sociology of the Mobile Phones*. Retrieved from http://geser.net/mobile/t_geser2.pdf

Glaser, B. G., & Strauss, A. L. (2009). *The discovery of grounded theory: Strategies for qualitative research*. Piscataway, NJ: Transaction.

Gray, N. J., Klein, J. D., Noyce, P. R., Sesselberg, T. S., & Cantrill, J. A. (2005). Health information-seeking behaviour in adolescence: The place of the Internet. *Social Science & Medicine*, *60*(7), 1467–1478.

Greenhow, C., & Lewin, C. (2016). Social media and education: Reconceptualizing the boundaries of formal and informal learning. *Learning, Media and Technology*, *41*(1), 6–30.

GSMA & NTT Docomo. (2012). *Children's use of mobile phones: An international comparison 2012*. Japan. Retrieved from http://www.gsma.com/publicpolicy/wp-content/uploads/2012/03/GSMA_ChildrensMobilePhones2012WEB.pdf?utm_source=Mobile+Learning+Week+2013_v3_CfP&utm_campaign=8885b82361-UNESCO_Mobile_Learning3_28_2013&utm_medium=email

Hargittai, E. (2001). Second-level digital divide: Mapping differences in people's online skills. Retrieved from https://arxiv.org/ftp/cs/papers/0109/0109068.pdf

Hargittai, E., & Hinnant, A. (2008). Digital inequality differences in young adults' use of the Internet. *Communication Research, 35*(5), 602–621.

Hargittai, E., & Jennrich, K. (2016). The online participation divide. In M. Lloyd & L. Frieldand (Eds.), *The communication crisis in America and how to fix it* (pp. 199–213). London, England: Palgrave Macmillan.

Horrigan, J. (2016b). *Lifelong learning and technology.* Washington, DC: Pew Research Center. Retrieved from http://www.pewinternet.org/2016/03/22/lifelong-learning-and-technology/

Horrigan, J., & Duggan, M. (2015). *Home broadband 2015.* Washington, DC: Pew Research Center. Retrieved from http://www.pewinternet.org/2015/12/21/home-broadband-2015/

Holden, C., Dikkers, S., Martin, J., & Litts, B. (2015). *Mobile media learning: Innovation and inspiration.* Retrieved from http://press.etc.cmu.edu/files/Mobile-Media-Learning_Holden-Dikkers-Martin-Litts-etal-web.pdf

Ito, M., Baumer, S., Bittanti, M., boyd, d., Cody, R., Herr-Stephenson, B., & Horst H. et al. (2007/2010). *Hanging out, messing around, and geeking out: Kids, living and learning with new media.* Cambridge, MA: MIT Press.

Ito M., Gutiérrez K., Livingstone S., Penuel W., Rhodes J., Salen K., ... Watkins, C. (2013). *Connected learning: An agenda for research and design.* Irvine, CA: The Digital Media and Learning Research Hub Reports on Connected Learning. Retrieved from http://dmlhub.net/publications/connected-learning-agenda-for-research-and-design/

Jenkins, H. (2006). *Fans, bloggers, and gamers: Exploring participatory culture.* New York, NY: NYU Press.

Kennedy, B. L., Brinegar, K., Hurd, E., & Harrison, L. (2016). Synthesizing middle grades research on cultural responsiveness: The importance of a shared conceptual framework. *Middle Grades Review, 2*(3), 2.

Klopfer, E. (2008). *Augmented learning: Research and design of mobile educational games.* Cambridge, MA: The MIT Press.

Kolb, L. (2008). *Toys to tools: Connecting student cell phones to education.* Arlington, VA: International Society for Technology in Education (ISTE).

Kuznekoff, J. H., & Titsworth, S. (2013). The impact of mobile phone usage on student learning. *Communication Education, 62*(3), 233-252.

Leinen, S. (2016). *Mobile lifestyles: The homeless and their cell phones.* Regent University. Retrieved from http://search.proquest.com.scsu.idm.oclc.org/docview/1870803283?pq-origsite=gscholar

Lenhart, A. (2009, April). Teens and mobile phones over the past five years: Pew Internet looks back. *Pew Report,* 1–12. Retrieved from http://www.pewinternet.com/~/media/Files/Reports/2009/PIPTeensandMobilePhonesData Memo.pdf

Lenhart, A. (2015, April). Teens, social media and technology overview 2015: Smartphones facilitate shifts in communication landscape for teens. *Pew Research Center,* Retrieved from http://www.pewinternet.org/2015/04/09/teens-social-media-technology-2015/

Lenhart, A., Arafeh, S., Smith, A., & Rankin Macgill, A. (2008). *Writing, technology and teens.* Washington, DC. Retrieved from http://en.scientificcommons.org/51860817%5Cnhttp://www.ncbi.nlm.nih.gov/pubmed/14744227%5Cnhttp://

www.fya.org.au/wp-content/uploads/2010/07/The-Benefits-of-Social-Networking-Services.pdf

Lenhart, A., Ling, R., Campbell, S., & K. Purcell, K. (2010). Teens and mobile phones. *Pew Internet & American Life Project*. Retrieved from http://www.pewinternet.org/Reports/2010/Teens-and-Mobile-Phones.aspx

Lesitaokana, W. O. (2014). "It's my mobile:" Adoption, consumption and meanings of the mobile phone among urban youth in Botswana—An ethnographic study. Retrieved from https://www120.secure.griffith.edu.au/rch/file/b73dab7c-feaf-4f1c-9b0e-8955079eb980/1/Lesitaokana_2014_01Abstract.pdf

Levin, D., & Arafeh, S. (2002). *The digital disconnect: The widening gap between Internet-savvy students and their schools*. Washington, DC: Pew Research Center and American Institute for Research.

Levinson, A, Siyahhan, S., Pressey, B., & Taylor, K.H. (2015). *Diverse families and media: Using research to inspire design/A casebook and design guide*. A report of the Families and Media Project. New York, NY: The Joan Ganz Cooney Center at Sesame Workshop.

Lim, S. S. (2013). On mobile communication and youth "deviance": Beyond moral, media and mobile panics. *Mobile Media and Communication, 1*(1), 96–101.

Ling, R. S. (2004). *The mobile connection: The cell phone's impact on society*. San Francisco, CA: Morgan Kaufmann.

Ling, R. (2005). Mobile communications vis-à-vis teen emancipation, peer group integration and deviance. In R. Harper, A. Taylor, & L. Palen (Eds.), *The inside text: Social perspectives on SMS in the mobile age* (pp. 175–189). London, England: Kluwer.

Ling, R. (2012). *Taken for grantedness: The embedding of mobile communication into society*. Cambridge, MA: MIT Press.

Ling, R., & Haddon, L. (2008). Children, youth and the mobile phone. In K. Dortner & L. Livingstone (Eds.), *The international handbook of children, media, and culture* (pp. 137–151). London, England: SAGE.

MEE Productions. (2008). A look inside urban youth culture. *Urban Trends Collection, 18*(1), 1–10.

Mertens, S. B., & Anfara, V. A. (2007). *The young adolescent and the middle school*. Charlotte, NC: Information Age Publishing.

Meyers, W. D. (1999). *Monster.* New York, NY: Amistad/HarperCollins.

Pain, R., Grundy, S., Gill, S., Towner, E., Sparks, G., & Hughes, K. (2005). "So long as I take my mobile": Mobile phones, urban life and geographies of young people's safety. *International Journal of Urban and Regional Research, 29*(4), 814–830.

Park, Y. J. (2014). My whole world's in my palm! The second-level divide of teenagers' mobile use and skill. *New Media & Society*. Retrieved from http://journals.sagepub.com/doi/abs/10.1177/1461444813520302?rss=1

Portus, L. (2005). World making: Mobile phone discourses among select urban poor married couples. *Plaridel, 2*(1), 101–122.

Poushter, J. (2016a). *Smartphone ownership and Internet usage continues to climb in emerging economies*. Washington, DC: Pew Research Center. Retrieved from http://www.pewglobal.org/2016/02/22/smartphone-ownership-and-internet-usage-continues-to-climb-in-emerging-economies/

Rainie, L., & Hitlin, P. (2005). The Internet at school. *Pew Internet and American Life Project.* Retrieved from http://www.pewinternet.org/Reports/2005/The-Internet-at-School/Data-Memo.aspx

Rice, E., Lee, A., & Taitt, S. (2011). Cell phone use among homeless youth: Potential for new health interventions and research. *Journal of Urban Health, 88*(6), 1175–1182. Retrieved from http://doi.org/10.1007/s11524-011-9624-z

Rogoff, B., Callanan, M., Gutiérrez, K. D., & Erickson, F. (2016). The organization of informal learning. *Review of Research in Education, 40*(1), 356–401.

Shenton, A. K., & Dixon, P. (2004). The development of young people's information- seeking behaviour. *Library and Information Research, 28*(90), 25–34.

Squire, K. (2011). *Video games and learning: Teaching and participatory culture in the digital age*. New York, NY: Teachers College Press.

Strauss, A., & Corbin, J. (1998). *Basics of qualitative research: Grounded theory procedures and techniques*. London, England: Sage.

Thulin, E., & Vilhelmson, B. (2009). Mobile phones: Transforming the everyday social communication practice of urban youth. In S. Campbell & R. Ling (Eds.), *The reconstruction of space and time: Mobile communication practices* (pp. 137–158). London, England: Routledge.

Wigfield, A., Lutz, S., & Wagner, L. (2005). Early adolescents' development across the middle school years: Implications for school counselors. *ASCA: Professional School Counseling, (9)*2, 112–119.

PART II

CONNECTING HISTORY AND COMMUNITY: LOCATION-BASED MOBILE LEARNING

CHAPTER 5

STEMlandia

Using Mobile Technology to Get 'Em Outside

James L. Larsen
Jodi Asbell-Clarke
Barbara MacEachern
Elizabeth Rowe
Educational Gaming Environments (EdGE) at TERC

As digital technologies and "screen time" grow increasingly pervasive in today's society, there is a fear that people are becoming disconnected from their natural surroundings and this could have detrimental impacts on learning and lifestyles. Ironically, mobile digital technologies may also get people reconnecting with the natural world. Holden, Dikkers, Martin, and Litt (2015) suggest that "mobile enables us to make learning come alive in new ways" (p. 13).

This chapter focuses on how location-based mobile learning technologies, such as smartphones, offer a new platform to create engaging and physically active experiences for learners of all ages, integrating STEM content related to their own natural environment. An example called *STEMlandia* illustrates how the geocaching and authoring tools on mobile phones can enable communities to develop their own narratives and adventures. We conclude with a discussion of future research possibilities to promote outdoor digital learning adventures using mobile technologies.

Mobile Learning: Perspectives on Practice and Policy
pp. 95–117

THEORETICAL AND PRACTICAL GROUNDING FOR DIGITAL OUTDOOR LEARNING ADVENTURES

As science educators and game designers, we are both enamored with, and dismayed by, the invasion of technology into today's society. While accepting that today's youth are immersed within a screen-based culture that travels with them in their pocket, we seek avenues to use that digital influence to re-engage youth and adults with their physical surroundings. We focus this chapter on educational mobile applications that, in the words of Holden et al. (2015, p. 13), "push back on the traditional boundaries of School to connect learning to the real world."

Reconnecting With the Natural World

With over 90% of youth playing digital games (Entertainment Software Association, 2013) and mobile devices becoming an increasingly integral part of children's lives, there is concern that children are missing out on learning and health benefits associated with being outside. Louv (2009) coined the term "nature deficit" to refer to the behavior problems children experience as a result of spending less time outdoors. The benefits of nature have been documented in terms of cognition (Dieser & Bogner, 2016); identity formation (Williams & Chawla, 2016); coping with stress and adversity (Corraliza, Collade, & Bethelmy, 2012); and the well-being of adolescent girls (Wiens & Kingäs, 2016).

While using the Internet to get people outside seems paradoxical, mobile technologies such as smartphones and wearable technologies present unique opportunities to reengage learners with their local natural environment. Direct exposure and experience with nature is important for children in terms of the development of biological knowledge. Cheng and Monroe (2012) showed that enjoyment of nature, empathy for other living things, and how youth viewed their place in, and responsibility for, the natural world was positively related to their levels of participation in nature-based activities and ultimately influence their interest in environmentally friendly behaviors. Sampson (2015) underscores the impact of the disconnect between kids and nature to their emotional and intellectual health, and how "denaturing childhood" is closely defined with a lack of understanding of sustainability both for human society and the ecosystems that comprise local and global environments.

Technology is often blamed for the increase in the public's time spent indoors, enraptured by screens texting, playing games, and plugged-in (Louv, 2013; Sampson, 2015). These are seen as huge time sinks that draw children (and adults) away from the natural world, but can also be seen as an affordance, or a tool to work with (Louv, 2013; Sampson, 2015). Louv introduces the notion of the "hybrid mind" to describe the ability of today's connected generation to live

> simultaneously in both the digital and physical worlds, using computers to maximize our powers to process intellectual data and natural environments to ignite our senses and accelerate our ability to learn and feel-combining the resurfaced 'primitive' powers of our ancestors with the digital speed of our teenagers. (Louv 2013, p. 1)

Sampson notes that although some psychologists equate our dependence on looking at screens to an "eating disorder," technology is now an "essential part of our daily diet" and must be considered.

The ways youth use mobile technology varies throughout society and should be considered with a culturally relevant lens, emphasizing the need and value for participants to contribute their personal idiosyncratic interests and identifications (e.g., Ladson-Billings, 1994; Moll & Greenberg, 1990). In our digitally driven society, youth identity formation may be even more greatly tied to their experiences with media rather than ethnic, institutional, or geographic affiliations, although those dimensions of self and identity continue to be relevant (Dolby & Rizvi, 2007; McCarthy, Giardina, Harewood, & Park, 2003). Engaging youth in digital storytelling about their own environs joined with the benefits of nature may be key to using technology outdoors, and this combination is uniquely afforded by mobile devices such as phones.

Leveraging Mobile Technologies

Mobile technologies are exploding in availability and may even help narrow the digital divide among different groups of learners (Shuler, 2009). Recently *Pokémon Go* (Niantic, Inc., 2017) exploded on the scene providing an example of an augmented reality game using mobile phones, but educational groups, particularly in citizen science, have been exploring mobile apps for a while.

Publicly available apps of this sort range from bird identification applications like Cornell Lab of Ornithology's Merlin (2017), to data logging and citizen science related mobile apps such as eBird (2017), and USA National Phenology Network's Natures Notebook (2017). When these types of tools also allow for authoring experiences, often in the form of games, they can offer a project-based approach for learners to practice 21st century skills like computational thinking as well as connecting classroom concepts and Next Generation Science Standards to their own

STEM (science, technology, engineering, and mathematics) adventures to share with others.

Two interesting geolocation apps for learning that use mobile phones for augmented reality are ARIS and TaleBlazer. They are both available free to educators and enable location-based storytelling. ARIS is an open-source program with a focus on creating mobile games, tours, or interactive stories. It consists of an online authoring tool and a smartphone app to run projects, and uses GPS and QR codes to help find locations. To make an adventure more interactive, an ARIS author can create characters to find and interview, items to pick up or drop, and place media to watch or interact—all tied to specific locations in the physical environment. Examples of mobile learning experiences created with ARIS include: Steel-Mining and Smelting in 1867, designed to be played on the streets of Madison, WI, and Mentira, designed with interactions around a historical novel that mix fact and fiction to drive the story while engaging learners in developing their Spanish speaking language skills to solve a murder mystery in a local neighborhood (ARIS Featured Projects, 2016).

TaleBlazer, like ARIS, is free to educators. It also uses a browser-based editor for creating projects, with completed projects accessed by an app on GPS-enabled smartphones. It allows for engaging user experiences where players can interact with characters and objects, and complete tasks tied to locations. TaleBlazer has a full set of resources to help facilitate using in formal and informal learning spaces, and with its visual blocks-based scripting language, it can be used to help teach computational thinking and coding skills while linking STEM learning and exploration in the outside world.

Examples of TaleBlazer games that utilize outdoor sites include several made at the University of Utah's Red Butte Garden (Red Butte Garden, 2016). Designers created five free-choice games ranging from Lady Nightshade and the Invaders, a game where players explore Red Butte's Water Pavilion Garden and learn about native plants and invasive species; to Water Games, situated in the Children's Garden, where players learn about water use efficiency and conservation; to Garden Fresh, which explores several parts of the garden, including the herb and medicinal gardens, to help a restaurant select the best ingredients as players take on the role of assistant chef and try to earn a 5-star rating; to Captain Bonneville, which takes place in the Garden's natural area, where players have to navigate and use geographic knowledge to explore with the added bonus of exercise as they have to cover a large area for the win.

There are a variety of other location-based and authoring tools available on mobile phones; however, ARIS and TaleBlazer have established communities of formal and informal educators, resources to help create learning adventures, are grounded in research communities, and are free.

Another promising form of location-based adventure using phones is geocaching (Schneider, Silverberg, & Chavez, 2011). Geocaching exploits handheld global positioning systems (GPS) to find hidden treasures geolocated around the world. Geocachers typically create their own hidden caches to share with others. Caches can range from a simple small container with a log report, to an intricate puzzle, to a series of caches strung together to create a multicache series at an interesting site. There are millions of active geocachers worldwide and they are predominantly networked through one main provider—Geocaching.com from Groundspeak, Inc. Anyone 13 or older can have a free account and engage in geocaching with GPS or smartphone technology.

Geocaching, first introduced in 2000, now has over 10 million registered geocachers searching for nearly 3 million active caches around the world (Groundspeak, 2016). These GPS-integrated treasure hunts often contain rich content about locations where caches are hidden (Dunlap, 2013; Groundspeak, 2016; Lewis, 2006; Neustaedter, Tang, & Judge, 2013; O'Hara, 2008).

Geocaching has been demonstrated to be highly motivating for participants (Gram-Hansen, 2009); however, it has been underutilized for educational purposes (König & Bernsen, 2014; Jones, Scanlon, & Clough, 2013) and for youth (Telaar, Krüger, & Schoning, 2014). One exception is EarthCaching. The Geological Society of America, in partnership with Groundspeak, has been creating and sharing EarthCaches, location-based activities tied to local geology. EarthCaches help connect and educate the public about various geoscience concepts like rock types, fossils, volcanoes, and unique geological features. Above all, EarthCaches are meant to be educational with a set of guidelines distinctly tied to the geosciences (EarthCache, 2017).

Pokémon Go: A Game for the Hybrid Mind Generation

With the launch of the AR gaming app Pokémon Go, the real world was quickly flooded with a mobile learning designer's dream: gamers who were once sedentary and stuck inside on screens were now outside exercising or experiencing nature for extended periods for the first time.

For some designers, it was the user interface that caught their attention, as they rethought how they might modify their existing apps to be more interactive—like capturing characters by swiping or engaging in Pokémon gym battles. For others, it was the ability to attach the gaming experience to anyplace and draw people to locations of interest as a way to get them outside exercising in nature. Bryan Roth, one of the cofounders of Geocaching.com describes Pokémon Go as a "unique phenomenon" that leveraged Nintendo's intellectual property with a "well-built location-based gaming platform" and "engaged a significant worldwide population of smartphone users in just a few days" (Schlosser, 2016, p. 1). Roth appreciates the fact that Pokémon Go was similar to geocaching in that it got "people outside having fun and away from their couches." Roth further noted how geocaching developers are also looking at how to turn every location into an adventure, allowing discovery to hide just around the next corner in every city and town. This may include a virtual component to geocaching and using emerging technologies like radio frequency identification, iBeacons and other location-aware technologies.

Practitioners are figuring out ways to use Pokémon Go for teaching data literacy, infographics, mapping and other useful 21st century skills (La Flamme & Gauthier, 2016; Schrock, 2016). This bodes well for our vision of STEMlandia adventures, especially as new tools are developed that engage more educators to work with learners to push the boundaries beyond the horizon. Pokémon Go showed that a mobile experience that leverages an audience of critical mass, provides an engaging storyline and user interface and taps into local sites, has the ability to change our connection to the real world.

Pokémon Go also reminds us that engaging learners in mobile games that take people outside require attention to the dangers to which players might be exposed. STEMlandia designers must remember that many things in nature can bite, stick, or cause harm, so it is important to be aware of surroundings when creating STEMcaches. In turn, as noted in many stories about Pokémon Go, outside explorations include the possibility of others that might do one harm. It would be wise for the mobile learning community to address such dangers with guidelines for thinking about, and safeguarding, outside mobile learning. As we re-engage learners with nature, we must be aware of the possibilities, but also the limits and challenges presented by both the technology we use and the environments we explore.

STEMLANDIA: OUTDOOR DIGITAL ADVENTURES

Informed by experiences with ARIS, TaleBlazer, EarthCaching and our research on free-choice gaming (Asbell-Clarke et al., 2012), our team began piloting STEMlandia activities to reengage the public with STEM-infused stories in their natural environs (STEMlandia, 2014). Geocaching was one of the first public uses of geolocation tools such as GPS, which was arguably the parent of the mobile technology revolution. Enabling an accurate reading of position in a pocket or wrist device launched a movement of families and outdoor enthusiasts who wanted to join in a globally connected, mobile-technology aided, hide and seek game.

STEMlandia leverages the phenomena of geocaching, and ready-made geolocation tools to create STEMcaches—a series of geocaches telling a STEM story about the local outdoor environment. While TaleBlazer and ARIS were considered as authoring tools, we ultimately went with a tool under development by Groundspeak called the Geocaching Adventures Platform (hereafter referred to as GAP). Although this is a proprietary tool used to create temporary Lab Caches and location-based geocaching adventures for special events around the world, Groundspeak was interested in leveraging it for educational use. STEMlandia offered an opportunity that allowed our team and Groundspeak to collaborate on studying how their tool might best be used in education.

Teaming with Groundspeak also provided a large, established community (with over 330,000 active users in the United States) attuned to exploring the world through a game of hiding and seeking physical and virtual caches. Groundspeak helped get the word out through its blog and newsletter blasts which helped draw Geocachers to the Arboretum pilot as well as provide local geocaching experts ready to work with students on future projects to develop new STEMlandia adventures. Eventually, as we build new STEMlandia adventures, they will be linked through the Groundspeak website and also published and made available via a permanent STEMlandia website.

Following the model of EarthCaches, we designed STEMcaches where geolocated puzzles are used to highlight a STEM concept related to a location. Each local environment is thought of as a STEM story waiting to be told and the STEMcaches reveal clues that work together to tell the story of the setting with special focus on the particular STEM issue determined appropriate.

The development of STEMlandia adventures is conducted using the GAP tool, which has a browser-based editor for creating adventures to share on GPS-enabled smartphone through a browser. Like ARIS and TaleBlazer, GAP uses GPS to identify and initiate interactions such as reading a narrative, viewing an introductory or follow-up video, linking to online support pages, and entering codes for finding and solving a puzzle for a cache. Although it is currently in beta, Groundspeak will make GAP available (through EdGE) to educators interested in creating their own location-based adventures. Unlike ARIS and TaleBlazer, GAP does not provide as large a variation of interactivity, such as coding nonplayer characters (NPCs) or virtual clues to pick-up or drop off, and, as a result, is much easier to use. This allows for more time focusing on story, STEM-cache creation, and location-driven observation and interaction, and less on learning the nuances of coding, game mechanics, and creating assets for driving gameplay.

The landscape for location-based learning is ripe and the players continue to evolve. Which tool is best depends on educational objectives, outreach goals, and the level of expertise and time available to develop a particular project. With the recent popularity of Pokémon Go (2017), one can expect a wave of new tools, or modifications to existing tools that reflect lessons learned.

BUILDING STEMCACHES: NATURE'S APPRENTICE

STEMlandia is intended to draw people outdoors and back to nature, noticing the natural world in their own (often ignored) local community environment. With this goal, we conducted the first pilot project of a STEMlandia mobile adventure called Nature's Apprentice in a public arboretum. This section describes this first pilot study.

Design and Implementation of Nature's Apprentice

In 2014, we conducted the first pilot study of STEMlandia at the Arnold Arboretum of Harvard University. Nature's Apprentice consisted of a series of 10 STEMcaches distributed throughout the Arboretum over a period of approximately 2 months in fall 2014.

Arnold Arboretum is in Jamaica Plain, an urban and low socioeconomic status area of Boston, and hosts 250,000 visitors per year. With its director of public programs, we identified and codeveloped a series of ten physical geocaches, the STEMcaches, located near plants, like Typha (cattail), Syringa (lilac), and Sequoia (Redwood) throughout the Arboretum highlighting STEM-related concepts. Typha were chosen to highlight their internal structure that allows them to deal with forces of birds landing on them as well as the wind; Syringa to tell a story on research using the lilac genome to track their spread around the world; and Sequoia to tell a story of fire-retardant bark and the importance of fire to their ecology. The Nature's Apprentice adventure was designed so that it could be played by an individual or group using the downloadable geocaching app on their phones. Puzzles tied to STEMcaches ranged from easy (upper elementary age-appropriate) to difficult (middle school and higher age-appropriate). The target audience, beyond geocachers, was the general public visiting the Arnold Arboretum.

Our team met with the educational staff of the Arboretum to plan a series of sites that:

- had a good "STEM story" or connection;
- introduced the diversity of plants located on the grounds;
- lent themselves to a puzzle or observation that linked STEM content to a solution;
- let players explore the grounds without excess physical effort;
- felt like an adventure—not too easy, not too hard;
- was extendable with hands-on experiential activity in future iterations;
- was generic enough to serve as a model for similar adventures at other locations.

Nature's Apprentice participants used their own GPS-enabled smartphones and a series of hints and puzzles to find each STEMcache. They used location-specific clues to solve puzzles and decipher a code that was entered via the smartphone in order to get credit for finding each cache. The difficulty levels varied among the caches to provide a challenge for all ages, science backgrounds, and varied geocaching experiences.

As we developed STEMlandia it was important not to create just a "tour" of the Arboretum as there were already brochures available that guide visitors to specific locations. Background information needed to solve the puzzles was delivered via the Groundspeak app so the digital experience was self-contained to a commonly available app, and this was supplemented with a paper "passport" that provided a map with general locations for each STEMcache.

We followed the model for mobile learning design put forth by Holden et al. (2015, p. 27) that is illustrated below with examples from STEMlandia's Nature's Apprentice project.

Identify a Connection

We wanted the *STEMlandia* adventure to take the learner to a specific place—a place they might normally have walked by—and explore a deeper connection to STEM concepts related to the plant or location.

In one example, the Plant Barcodes STEMcache in Nature's Apprentice, the puzzle was tied to the vast lilac collection along the Arboretum's Bussey Hill Road and focused on a Finnish researcher's work to identify genetic markers in the lilac genome to see what they told her about the relatedness of lilacs around the world. The puzzle presented on the Passport read: *Take the simplest code found in all living things, arrange and replace first to last, then drop the naught.* Additional clues on the passport for this included pictures of Adenine and Thymine, two examples of nucleobases, additional background on the support website, and an activity to collect DNA from strawberries that players could do at home to help them think about and solve the clue. This turned out to be one of the most difficult puzzles for players to solve. In another example, Newton <3 Biology, the puzzle made a connection between the form and function of plants by asking the learner to consider a Dawn Redwood and a Cattail and the forces they had to deal with in order to transport water or withstand the wind without breaking.

Given the limited time and resources available at each stop, we had to carefully script text to deliver enough informational content to allow the player to observe and think about what they saw at each location as well as provide a puzzle that required deciphering a clue to discover a code to enter into the smartphone app and get credit for finding a cache and solving its puzzle. The amount of scaffolding needed to be sufficient for a broad array of learners in terms of age, reading ability, and science background. For example, some puzzles were easily solved by finding a locked cache and discovering a number code on one of the record labels found on nearby trees. Others, like the Plant Barcodes cache described previously, were intentionally hard to provide a challenge to those who appreciate more difficult puzzles. Puzzles used graphical elements and tied into the "props" availed by the environment as much as possible—like observing leaf structure at the Fitness of Leaves cache, or geometric shape for packing seeds in a pinecone at the CryptoBotanica cache.

Identify Desired Results

Our desired results were to leverage the public's fascination with mobile technology to make a connection to locations and related STEM concepts found outside of a textbook, nature show on television, or the Internet. As a pilot study, this research focused on the design features for successful implementation and engagement of geolocation mobile activities with a broad and diverse audience. While STEMlandia was designed to support science learning, it was beyond the scope of this project to rigorously measure science learning by the participants.

By selecting locations that had STEM-related stories, we focused on meaningful content that could be discovered through fun, mobile-facilitated experiences. Since we could not control the age of the general public taking part in STEMlandia, we had to carefully balance the locations, contents and puzzles for each cache so that some would be easy enough to do by families with elementary-aged learners and some hard enough to challenge seasoned geocachers.

In addition, since we were researching the STEMcache approach, we wanted to attract as diverse an audience as possible. We reached out to the geocaching community who were already using the mobile app, and also to regular visitors to the Arboretum (through their newsletter) and the greater Cambridge community who might not know about geocaching or the Arboretum. To accomplish outreach beyond geocaching and Arboretum communities, *STEMlandia* was launched at the National Science Teachers Association national conference held in Boston and also listed as part of the annual Cambridge Science Festival.

To examine how features of the geolocation mobile experience aided in obtaining these desired results, we conducted pilot research on the design and implementation of the STEMlandia experience.

RESEARCH ON NATURE'S APPRENTICE

The goal of STEMlandia adventures is to use mobile technologies to engage participants in physically active learning about their natural environment. In the pilot STEMlandia project called Nature's Apprentice, players were to use the geocaching app on their phones to find as many STEMcaches as they could, over roughly a 1.5 mile walk (on the most efficient route), applying STEM content and concepts to solve each new puzzle along the way.

We used two forms of evidence to examine engagement and learning in the pilot study of Nature's Apprentice. The GAP app running on a player's smartphone kept track of location and time data tied to entering a correct answer or code deciphered from STEM content specific to each STEMcache puzzle. Players were also given a paper "Passport" where they could fill out their answers to the puzzles and drop them at the Arboretum visitors' center. Our research focused on the following questions:

1. What drew participants to the Nature's Apprentice experience?
2. How did participants make use of the Nature's Apprentice tools and resources?
3. What did participants find worthwhile in the Nature's Apprentice experience? What did they find challenging or frustrating?
4. How likely are they to participate again? Encourage others to participate? Why?

Sample and Data Sources

During the 8-week pilot study, 175 unique groups used their smartphones to log into the URL used for Nature's Apprentice, which likely represents about 300 participants as many participated in groups, including families. Researchers examined GAP log data and Google analytics, along with surveys ($N = 41$) and informal interviews conducted with participants during their experience.

About 68% of survey takers were between the ages of 35 and 54. Two of the participants looked for the caches on their own, but most participated with friends and family members. About 32% of the survey takers had children with them whose ages ranged from 6 to 16. On average, the survey takers reported that two or three members of their exploring group used the Geocaching Labs STEMlandia Web app with only 41% of the groups having a sole individual using the app. This suggests design of mobile experiences that support collaborative learning with roles for multiple partners (e.g., searching, data collection, data entry) and research methods that do not presume tracking of individual users.

The 41 participants who completed the survey had been members of Geocaching.com for an average of 4.5 years at the time of the STEMlandia adventure, which is not significantly different from the average of 4.1 years for those not completing the survey but whose activity was recorded by the Geocaching app (t test, $p = .09$). The survey takers did, however, differ significantly from the rest of the participants in regard to how many total caches they found. On average the survey takers found 9.6 total caches, which is significantly higher than the nonsurvey takers, who found an average of 6 total caches (t test, $p < .001$). Their high level of engagement and persistence with the adventure may have been paired with their identity as seasoned geocachers and/or a desire to provide feedback to the cache designers.

Research Question 1: What Drew Participants to the Nature's Apprentice experience?

Among the 41 survey respondents, 59% indicated hearing about STEMlandia on Geocaching.com and 93% indicated being a regular geocacher. When describing their motivations for participating in this geocaching adventure, 41% of survey respondents referenced participating in order to be part of a "lab cache," which is defined on the Geocaching.com blog as "a new and rare geocache type ... being tested by Geocaching Labs." For the participants who completed the survey, who on average found 9.6 of the 10 caches, it seems that the geocaching aspect of this experience was the biggest initial draw to participate. This was confirmed by analysis of the geocaching accounts. Only 35 participants (21%) used geocaching accounts that were less than 60 days old, which is the duration of the Nature's Apprentice experience. The other 79% of participants had existing geocaching accounts.

None of the participants who filled in the survey were members at the Arnold Arboretum and only 36% had been to the Arboretum before. *Nature's Apprentice* was an impetus for most participants to visit and explore the Arnold Arboretum for the first time.

Research Question 2: How Did Participants Make Use of the Nature's Apprentice Tools and Resources?

The 10 caches varied in difficulty; however, most players completed most of the caches. Table 5.1 shows a list of the caches with the number of participants finding each one. A few of the survey takers mentioned the Plant Barcodes cache as being especially difficult, the cache found by the fewest number of participants (79).

As a proxy for experience with geocaching, we examined how participation varied with age of the Geocaching.com account. Participants with older geocaching usernames, who presumably have more experience and knowledge of geocaching, did not necessarily find higher numbers of geocaches.

All survey respondents claimed that they referred to maps of the Arboretum and read signs posted in the Arboretum at least once, while 93% said they used the information in the Geocaching Lab Nature's Apprentice Web app at least once, making this a primarily mobile-based experience. About 85% of participants used the geographical coordinates the app provided at least once, with 66% of survey takers indicating that they

Table 5.1. Number of Participants Finding Each STEMcache in Nature's Apprentice

Cache Name	*Number of Participant*
Fitness of Leaves	141
Firefighters	132
Newton <3 Biology	128
Chemical Warfare	127
Open the Door	120
Citizen Science	116
CryptoBotanica	111
Sum of All Parts	94
STEMlandia Bug Launch	82
Plant Barcodes	79

used both the info in the app and the coordinates it provided more than two times during the adventure. One survey taker commented that the coordinates were often not necessary to solve the caches, which may be why not all participants used the coordinates. This was probably the case for three reasons. First, experienced geocachers develop what they call a "geosense" and take pride in looking at a location and figuring out where a cache may be hidden. Second, a more likely possibility is that the participants were at the Arboretum during the Cambridge Science Festival when there were so many taking part you could follow others to the locations and not depend on coordinates. Lastly, since we were designing for a diverse audience, some caches were designed to be quite easy to find to allow younger or less experienced geocachers the joy of finding and solving a puzzle for a cache.

About 75% of the participants utilized links on the Nature's Apprentice support website at least once (presumably on their phone) and 66% of survey takers spoke with Arboretum staff at least once. Only 4 of the survey takers mentioned using the Arnold Arboretum's own Arb Explorer app, which was not integrated into the Nature's Apprentice experience. Overall, Arboretum resources and Nature's Apprentice-specific resources were widely used among participants at least once on their Nature's Apprentice adventure through the Arboretum. The mobile app, in and of itself, was not sufficient to ensure participation, as the Nature's Apprentice app got far more participation than the Arb Explorer app.

Research Question 3: What Did Participants Find Worthwhile in the Nature's Apprentice Experience? What Did They Find Challenging or Frustrating?

All 41 of the survey respondents indicated that it was highly enjoyable to spend time outside during Nature's Apprentice and 94% of them indicated that geocaching enhanced their experience at the Arboretum. About 90% found exploring the Arboretum highly enjoyable and 85% highly enjoyed finding the caches. Most participants highly enjoyed solving the puzzles to unlock the caches (77% of survey takers), but 6% only mildly enjoyed solving the puzzles or did not enjoy the puzzles at all. Those who did not enjoy solving the puzzles cited frustration with the challenging nature of some of the caches or issues with the app. Although some reported troubles with the app or that the caches were too difficult, overall the survey takers expressed that they enjoyed the experience.

When asked to rate the app, 76% of the survey takers gave it 4- or 5-stars. Most of these individuals found all 10 caches, which likely influenced their rating.

Research Question 4: How Likely Are They to Participate Again? Encourage Others to Participate? Why?

On a 5-point scale from very unlikely to very likely to participate, at least 95% of the survey takers indicated they are somewhat or very likely to participate in this type of geocaching adventure again and/or would recommend STEMlandia to people they know. This is not surprising given the positive feedback most of these individuals provided throughout the survey.

Over half of respondents indicated that it was unlikely they would return to find all the caches. Out of these survey takers, all but one individual had already found all 10 caches, so would likely not return because they had already completed the adventure. However, many of the respondents who answered that they *were* likely to return to find all the caches had already found all 10. Perhaps this was an indicator they wished there were more caches to find, or perhaps they intended to return to participate with a different group of individuals.

About 80% of survey respondents expressed some interest in additional hands-on activities to be used in conjunction with the apps, but there was not much enthusiasm with only 51% indicating they would have been very likely to use hands-on activities at each cache, and 40% indicating very likely to use such activities at the visitor center. The proposal of adding hands-on activities to the mobile experience might be even more removed from traditional geocaching which is perhaps why these individuals are not necessarily as keen on this suggestion.

The pilot STEMlandia project, although a start that provided useful guidance for further work, presented unique challenges inherent in research conducted with mobile devices in natural settings. For example, too many restrictions, like defining specific groups to study, or intrusive data collection requirements, would likely detract from the enjoyment and, ultimately, the very connection of getting out into the nature we wish to study. As such, our current efforts are not only working to define the creation and sharing of mobile experiences in formal and informal learning spaces, but also looking at the best methodology, through multimodal and nonintrusive means, to capture the impact of STEMlandia stories on those experiencing them in the wild; especially in informal, open spaces.

NEXT STEPS: BUILD YOUR OWN STEMCACHES

While the Nature's Apprentice experience engaged the general public with STEM-related puzzles on their mobile devices designed by our team about their local environment, our designers did not feel that our goal of ownership and agency about local environments would be met with that design plan. We felt that we need to shift the focus of "who is doing the creative work of making" (Holden et al., 2015, p. 28) from educators to learners in order to have deep and sustained impact in a community. Thus, we strove to have learners be the "producers" of the content in their mobile learning experience rather than just "consumers."

We conducted several small STEMlandia outreach projects where we empowered teams of youth and educators in two urban after-school programs in the Boston area to build their own STEMlandia adventures using the mobile app to bring attention to features in local environments near their classrooms and local parks (Fresh Pond, Minuteman Rails to Trails Bikeway, Charles River). The teams learned about geocaching by finding a few local caches, then worked to develop some STEMcaches related to content tied to sports science, habitats, and ethnobiology.

We intended to launch and test their products as miniadventures for the rest of each program's youth to experience on their mobile devices, but the nature of the programs and limited funding did not allow for this phase of the project. In addition, the limited time did not allow for creating a storyline that we feel is crucial to the connectedness between a series of STEMcaches and the chosen setting.

This experience has, however, informed a proposed next phase of our work to define a model for learners in formal and informal settings to create STEMlandia adventures themselves. This will take place through our ongoing STEMlandia efforts, which are directed at working with educators to train them to work with their learners to design and implement STEMlandia Stories—a narrative-driven series of STEMcaches connected by puzzles with STEM-infused observations and solutions tied to local interests.

We are also finding inspiration from like-minded educators, like Dr. Polley, *Educacher Extraordinaire* (Geocaching Blog, 2015) who has students create caches as well as launch "Travel Bugs" to track as they travel around the world as part of Classroom Competitions and Academic Mystery Caches. Such initiatives do not utilize the mobile *GAP* tool, but would easily be integrated into future research plans.

We will continue to use the GAP tool and reach out to other educators using geocaching on mobile devices as a tool for STEM education, as well as explore other authoring tools and strategies (like Virtual Reality connections) to advance our research. Using these and emerging tools, we strive to continue development of STEMlandia experience based upon findings from this pilot research. The model we suggest for future STEMlandia experiences and other mobile outdoor adventures for STEM learning includes the following components:

1. Experience a STEMlandia Story—Working with local educators, create two or three STEMcache adventures for their learners to experience. Once created, these will be used to introduce learners to the concept of a STEMlandia Story. The STEMcache adventures will help both the educators and the learners to experience the components of a simple adventure as well as introduce them to the materials and tools needed to create a mobile learning STEMlandia Story.
2. Create a Community Asset Map—Working with educators and their students, explore the local community and construct a map of resources (people, places, and things) that will help in the design process and creation of relevant STEMlandia Stories.
3. Define Potential Stories and STEMcache Locations—Support educators and learners in using the Community Asset Map to focus on stories and locations of interest that can define the STEMcaches. For example, one school in the Southwest we are looking to work with has numerous locations along the Colorado River that will fit nicely with stories that are relevant to the community (e.g., water rights, agriculture, habitat restoration, impact of dams along the river, settling the West, and the importance of the river to the migrant community). The river is near a state park that has cataloged the military presence in the community as well as near a large agricultural area that depends on water from the river. Both areas have people, ranging from local farmers to park rangers to naturalists, who are eager to offer their expertise. In addition, the locations are along a corridor that brings a variety of people to the river for exercise (a rails-to-trails bike path) and for recreation (a wetland for birding and public park with various access points to the river).
4. Storyboard a Narrative and STEMcaches—After deciding on a narrative and potential STEMcache locations, learners will create a paper version to test using an iterative design process. This will allow learners to focus on content rather than getting mired in the development techniques for the mobile platform.
5. Create STEMlandia Story Assets—Once the narrative, locations, and STEMcaches are defined, paper tested, and approved, learners will create the physical and digital assets that comprise the mobile adventure. This will include a physical STEMcache, text along with video and other resources that would become part of the mobile STEMlandia Story.
6. Create Mobile STEMlandia Story (or Stories) for Testing—Using the appropriate authoring tool, a digital version to release to the intended audience will be created, tested and finalized for launch.

7. Revise and Define a Launch Event—A target launch date for the mobile learning experience will then be set based on the needs of the project. If it is only meant to be shared with a specific school community, the launch might be tied to a schoolwide event like the STEMlandia event, Canopy ReLeaf Geocaching Adventure, created by the Cincinnati Country Day School middle-schoolers. Their adventure was created and launched during a schoolwide effort to focus the community on the importance of trees to an ecosystem as well as the damage caused by the Emerald Ash Borer, an invasive species, on trees in the greater Cincinnati area.
8. Launch STEMlandia Story—Release the STEMlandia mobile experience to the target audience on the launch date and pay particular attention to how best to support and maintain the adventure for its duration. Initially, all adventures are meant to be temporary and to run for no more than a month to reduce the need for support.
9. Reflection and Long-Term strategies—Once an adventure has been completed, it might be archived, improved upon, or extended depending on feedback from the community and local needs.

DISCUSSION

STEMlandia, with the pilot project Nature's Apprentice, offers an innovative look at how mobile technologies can be used to improve STEM learning by reversing a trend toward the immersive and isolating screen experiences that exclude connections with the natural world. While digital outdoor adventures seemed novel and even marginal just last year, after the explosion of the mobile phone-based augmented reality game, Pokémon Go, this summer we expect to see many more blurred reality experiences that blend the digital and physical worlds we live in. This new modality offers novel avenues for design and research on learning.

Studying How People Learn in Digital Outdoor Experiences

Many place-based learning experiences have led the way for our research in mobile learning. Some, like "Mystery Trip" (Martin, 2012), included learners in the design process to help better connect learners to their community and environment. Another, "ParkQuest" (Dikkers, Rieder, & Solomon, 2015), used a storyline to encourage urban youth to use maps to navigate and explore as a way to discover plants and animal life in a state park. Both provide models and valuable lessons learned for future versions of STEMlandia, as well as for others to build upon with current AR tools. These examples, and other recent research, suggests that mobile location-based games have great potential to engage learners and promote learning (Klopfer 2008), but researchers have difficulty establishing assessments that yield evidence to make valid interpretations on student learning in such a rich domain (Squire & Jan, 2007).

One of the most powerful aspects of digital activities is that they leave digital footprints. Educational researchers now can collect digital log data such as a time-stamped record of a participant's clicks, swipes, opening and closing of apps—with mobile devices offering positional data as well. Researchers use data mining methods to analyze the data logs from these activities, much in the way Facebook and Amazon mine consumer behaviors. As learning scientists use these methods to build models of learning, educators and designers can use the resulting information about learning to create adaptive learning and teaching environments. Designers can create customized activities to keep each participant in their own zone of proximal development (Vygotsky, 1978) or state of flow (Csikszentmihalyi, 1990). Educators can use information about game-based learning to inform classroom teaching, adapting lessons to the knowledge base the game player is bringing to class.

Current research in game-based learning often employs educational data mining and learning analytics methods to discover which patterns of individual behavior in game data logs are associated with improvements in STEM cognition (Rowe, Asbell-Clarke, & Baker, 2015; Rowe, Baker, & Asbell-Clarke, Kasman, & Hawkins, 2014; Shute, Ventura, & Kim, 2013). Our research group has shown that data mining methods can infer implicit learning in digital games (Rowe et al., 2015), which in turn predicts improved learning on external assessments when the implicit learning is leveraged and bridged (facilitating transfer) to classroom learning (Rowe et al., 2015; Shute, Ventura, & Ke, 2015). To develop measures of implicit game-based learning, researchers coded extensive samples of video observation of gameplay to identify strategies players developed that appeared consistent with underlying science understandings. Concurrently, gameplay data logs were collected including each click by each player with a time stamp and record of the state of the game. The video observations and codes served as ground-truth to distill the game log data into features that could then be mined and modeled to find patterns in gameplay data consistently used by players. Different types of data analytics were used for each game, and even within a game, to detect and validate evidence of implicit learning.

To use data mining methods to measure learning in mobile experiences outdoors, the experiences must be designed such that the digital behavior maps with the intended learning activities. For outdoor, distributed, and social activities such as STEMlandia, the learning activities must be mapped across multiple participants, media, and settings. Learning and its mediating psychosocial factors are rarely studied together in educational research (Greeno, 1998), likely because of the difficulty of scalability and replicability when measuring these factors. However, the growing ubiquity of unobtrusive digital recorders (e.g., wearable technologies) and innovative methods of data analytics present new opportunities for measuring affective measures such as emotion and attention. Digital recorders also enable multimodal data collection to build multidimensional, dynamic models of learning. For example, we can imagine participants wearing PivotHead (or equivalent) glasses that record audio and video (what the wearer sees and hears) as well as GPS data. Cameras built in to mobile devices can use screen capture software to record the activity, gaze, and speech of the participant as well as their interactions with the device and salient apps. Gathering this type of multimodal data will enable thorough observation of the dynamic, distributed experience.

Understanding how mobile learning works in the wild is a key area for research as we open the door and help learners reengage and connect with nature. Location-based learning experiences show great potential for engaging learners with their community and environment, but the distributed nature of the activity adds challenges to research design. Fortunately, the next few years promise to add many new devices and methods that will enable rigorous measurement using data mining and other methods to study learning in outdoor digital adventures enjoyed by a broad and diverse audience.

AUTHOR NOTE

This work was partially funded by TERC and the U.S. National Science Foundation (NSF/EHR/DRL#1134919). The authors thank: their team at TERC for their work on STEMlandia, including Katie Stokinger, Erin Bardar, and Emily Kasman; Bryan Roth and the Groundspeak minions; and Daphne Minner for her support at the arboretum. Correspondence regarding this article should be sent to James L. Larsen, EdGE at TERC, 2067 Massachusetts Avenue, Cambridge, MA 02140 or e-mail to jamie_larsen@terc.edu.

REFERENCES

ARIS Featured Projects. (2016). Retrieved March 13, 2017, from http://arisgames.org/

Asbell-Clarke, J., Edwards, T., Larsen, J., Rowe, E., Sylvan, E., & Hewitt, J. (2012). Martian Boneyards: Scientific inquiry in an MMO Game. *International Journal of Game-Based Learning, 2*(1), 52–76.

Cheng, J. C., & Monroe, M. C. (2012). Connection to Nature Children's affective attitude toward Nature. *Environment and Behavior, 44*(1), 31–49.

Corraliza, J. A., Collado, S., & Bethelmy, L. (2012). Nature as a moderator of stress in urban children. *Procedia-Social and Behavioral Sciences, 38*, 253–263.

Csikszentmihalyi, M. (1990). *Flow: The psychology of optimal experience*. New York, NY: Harper & Row.

Dikkers, S., Rieder, R., & Solomon, T. (2015). ParkQuest: Mobile media learning as a large group activity. In C. Holden, S. Dikkers, J. Martin, & B. Litts (Ed.), *Mobile media learning: Innovation and inspiration* (pp. 211–218). Pittsburgh, PA: ETC Press.

Dieser, O., & Bogner, F. X. (2016). Young people's cognitive achievement as fostered by hands-on-centered environmental education. *Environmental Education Research, 22*(7), 943–957.

Dolby, N., & Rizvi, F. (2007). *Youth moves: Identities and education in global perspective*. London: Routledge.

Dunlap, M. A. (2013). *Science caching: Applying geocaching to mobile citizen science*. Calgary, Alberta, Canada: University of Calgary.

EarthCache. (2017). *EarthCache*. Retrieved March 13, 2017, from http://earthcache.org/

eBird. (2017). eBird: Birding in the 21st Century—Cornell Lab of Ornithology. Retrieved March 13, 2017, from http://ebird.org/content/ebird/

Entertainment Software Association. (2013). Game player data. *ESA Industry Facts*. Retrieved from http://www.theesa.com/facts/gameplayer.asp

Gram-Hansen, L. B. (2009). *Geocaching in a persuasive perspective*. Paper presented at the Proceedings of the 4th International Conference on Persuasive Technology, Claremont, CA.

Greeno, J. G. (1998). The situativity of knowing, learning, and research. *American Psychologist, 53*(1), 5–26.

Geocaching Blog. (2015). How Dr. Polly, Educacher Extraordinaire, Incorporates Geocaching in his Classroom. Retrieved March 13, 2017, from https://www.geocaching.com/blog/2015/10/how-dr-polley-educacher-extroadinare-incorporates-geocaching-in-his-classroom/

Groundspeak. (2016). Media FAQs. Retrieved March 13, 2017, from https://www.geocaching.com/press/faq.aspx

Holden, C., Dikkers, S., Martin, J., & Litts, B. (Eds.). (2015). *Mobile media learning: innovation and inspiration*. Pittsburgh, PA: ETC Press.

Jones, A. C., Scanlon, E., & Clough, G. (2013). Mobile learning: Two case studies of supporting inquiry learning in informal and semiformal settings. *Computers & Education, 61*, 21–32.

Klopfer, E. (2008). *Augmented learning: Research and design of mobile educational games*. Cambridge, MA: MIT Press.

König, A., & Bernsen, D. (2014). Mobile learning in history education. *Journal of Educational Media, Memory, and Society, 6*(1), 107–123.

La Flamme, A., & Gauthier, J. (2016). 14 ways to bring Pokémon Go to school. Retrieved March 13, 2017, from https://www.iste.org/explore/articleDetail?articleid=788

Ladson-Billings, G. (1994). *The Dreamkeepers*. San Francisco, CA: Jossey-Bass.

Lewis, G. B. (2006). *Earthcaching: Using geocaching to teach the public about the Earth.* Paper presented at the Philadelphia annual meeting of the Geological Society of America, Philadelphia, PA. Retrieved March 13, 2017, from https://gsa.confex. com/gsa/2006AM/finalprogram/abstract_105368. htm

Louv, R. (2009). Do our kids have nature-deficit disorder. *Educational Leadership, 67*(4), 24–30.

Louv, R. (2013). The hybrid mind: The more high-tech schools become, the more nature they need. *The New Nature Movement blog, Children & Nature Network.* Retrieved March 13, 2017, from https://www.childrenandnature.org/2015/03/30/the-hybrid-mind-the-more-high-tech-education-becomes-the-more-nature-our-children-need/

Martin, J. (2012). Mystery trip. In C. Holden, S. Dikkers, J. Martin, & B. Litts (Eds.), *Mobile media learning: Innovation and inspiration* (pp. 99–110). Pittsburgh, PA: ETC Press.

McCarthy, C., Giardina, M. D., Harewood, S. J., & Park, J. K. (2003). Contesting culture: Identity and curriculum dilemmas in the age of globalization, postcolonialism, and multiplicity. *Harvard Educational Review, 73*(3), 449–465.

Merlin. (2017). All About Birds—Cornell Lab of Ornithology. Retrieved March 13, 2017, from https://www.allaboutbirds.org/guide/Merlin/id

Moll, L., & Greenberg, J. (1990). Creating zones of possibilities: Combining social contexts for instruction. In L. C. Moll (Ed.), *Vygotsky and education: Instructional implications of sociohistorical psychology* (pp. 319–348). Cambridge, England: Cambridge University Press.

Nature's Notebook. (2017). Nature's Notebook. Retrieved January 22, 2017, from https://www.usanpn.org/natures_notebook

Neustaedter, C., Tang, A., & Judge, T. K. (2013). Creating scalable location-based games: Lessons from geocaching. *Personal and Ubiquitous Computing, 17*(2), 335–349.

Niantic, Inc. (2017). Pokémon Go. Retrieved January 22, 2017, from http://www.pokemongo.com/

O'Hara, K. (2008, April). *Understanding geocaching practices and motivations.* Paper presented at the Proceedings of the SIGCHI Conference on Human Factors in Computing Systems, Florence, Italy.

Pokémon Go. (2017). *Pokémon Go*. Niantic Inc. Retrieved from http://www.pokemongo.com/

Red Butte Garden. (2016). TaleBlazer Games. Retrieved March 13, 2017, from http://www.redbuttegarden.org/taleblazer

Rowe, E., Asbell-Clarke, J., & Baker, R. S. (2015). Serious games analytics to measure implicit science learning. In C. S. Loh, Y. Sheng, & D. Ifenthaler (Eds.), *Serious games analytics* (pp. 343–360). Cham, Switzerland: Springer International.

Rowe, E., Bardar, E., & Asbell-Clarke, J. (2015). Building bridges: Teachers leveraging game-based implicit science learning in physics classrooms. In D.

Russell & J. Laffey (Ed.), *Handbook of research on gaming trends in P–12 education* (pp. 442–468). Hershey, PA: IGI Global.

Rowe, E., Baker, R., Asbell-Clarke, J., Kasman, E., & Hawkins, W. (2014, July). Building automated detectors of gameplay strategies to measure implicit science learning. Poster presented at the Seventh international conference on Educational Data Mining Society, London, England.

Sampson, S. D. (2015). *How to raise a wild child: The art and science of falling in love with nature.* New York, NY: Houghton Mifflin Harcourt.

Schlosser, K. (2016). *Geocaching co-founder: Pokémon Go has been great for our business.* Retrieved March 13, 2017, from http://www.geekwire.com/2016/geocaching-co-founder-on-pokemon-go/

Schneider, I. E., Silverberg, K. E., & Chavez, D. (2011). Geocachers: Benefits sought and environmental attitudes. *LARNet-The Cyber Journal of Applied Leisure and Recreation Research, 14*(1).

Schrock, K. (2016). Pokémon Go in the classroom. Retrieved March 13, 2017, from http://blog.discoveryeducation.com/blog/2016/07/13/pokemongo/

Shuler, C. (2009). *Pockets of potential: Using mobile technologies to promote children's learning*. Retrieved March 13, 2017, from http://www .joanganzcooneycenter.org/wp-content/uploads/2010/03/pockets_of_potential

Shute, V. J., Ventura, M., & Ke, F. (2015). The power of play: The effects of Portal 2 and Lumosity on cognitive and non-cognitive skills. *Computers & Education, 80*, 58–67.

Shute, V. J., Ventura, M., & Kim, Y. J. (2013). Assessment and learning of qualitative physics in Newton's Playground. *The Journal of Educational Research, 106*, 423–430.

STEMLandia (2014). STEMlandia. Retrieved March 13, 2017, from http:// STEMlandia.com

Squire, K. D., & Jan, M. (2007). Mad City Mystery: Developing scientific argumentation skills with a place-based augmented reality game on handheld computers. *Journal of Science Education and Technology, 16*(1), 5–29.

Telaar, D., Krüger, A., & Schöning, J. (2014). A large-scale quantitative survey of the German geocaching community in 2007. *Advances in Human-Computer Interaction, 2014*, 7.

Vygotsky, L. S. (1978). *Mind in society: The development of higher psychological processes.* Cambridge, MA: Harvard University Press.

Wiens, V., Kyngäs, H., & Pölkki, T. (2016). The meaning of seasonal changes, nature, and animals for adolescent girls' wellbeing in northern Finland: A qualitative descriptive study. *International Journal of Qualitative Studies on Health and Well-Being, 11*, 10.3402/qhw.v11.30160. http://doi.org/10.3402/ qhw.v11.30160

Williams, C.C., & Chawla, L. (2016). Environmental identity formation in nonformal environmental education programs. *Environmental Education Research, 22*(7), 978–1001.

CHAPTER 6

COMMUNITY INQUIRY WITH MOBILE ASSET MAPPING

Ryan M. Rish
University at Buffalo (SUNY)

Aijuan Cun
University at Buffalo (SUNY)

Abigail Gloss
State University of New York at Fredonia

Merve Pamuk
University at Buffalo (SUNY)

This chapter reports on a case study of an adolescent participating in the Community Inquiry and Mobile Learning (CIML) pilot project in Western New York. Conducted in the summer of 2016, the 3-week pilot project was designed to engage middle and high school participants in mobile asset mapping as a form of community inquiry. A participant in the study, Adam (pseudonym), used a mobile application (i.e., ArcGIS Collector) to collect geographic, descriptive, and multimedia information about opportunities for playing sports in his community. Adam then presented this information on an interactive Story Map comparing the data he collected with map-based population statistics. Adam's Story Map documents the relationship

Mobile Learning: Perspectives on Practice and Policy
pp. 119–141

between opportunities for playing sports and the accessibility of these opportunities based on where people live in his community. This case study presents how Adam's mobile asset mapping helped him extend his knowledge of his community beyond his own personal experience growing up there. In this chapter, we present Adam's community inquiry project as a salient case of mobile learning.

COMMUNITY INQUIRY

Our approach to engaging adolescents in community inquiry stems from three arguments about supporting learning opportunities for youth. The first argument is that learning opportunities can be established to connect multiple places of learning. Our aim is to disrupt the notion that learning is limited to the attainment of isolated skills within classrooms during particular times of day and year. In our view, youth are engaged in complex mobilities and trajectories that establish relationships between the different physical places and virtual spaces they navigate when learning and establishing/maintaining social relationships. We define mobility as encompassing:

- physical mobility, the embodied movement from place to place;
- virtual mobility, the use of technology and media to build social connections across time and space; and
- educational mobility, the distribution of learning opportunities across people, tools, and learning environments (Leander, Phillips, & Taylor, 2010).

Therefore, we seek to understand not only how learning takes shape in different ways in distinct places (e.g., classroom, playground, corner store), but we are also interested in supporting learning across these places in terms of what adolescents bring along with them from place to place, as well as what learning is supported in those places.

The second argument is that learning opportunities should be oriented toward the lived experiences of people as they grapple with the social complexities that shape their lives (Bruce & Bishop, 2008). Our goal is to work with adolescents on social issues that they consider relevant to their past, present, and future lives. We argue that community inquiry as a learning opportunity should begin with an orientation toward issues that are relevant to youth and eventually build to include issues that impact others. Therefore, community inquiry is conceptualized not only as a learning opportunity for an individual, but rather as a learning opportunity that involves many people who share their lived experiences and collaboratively identify social issues that are collectively considered to be relevant. To support learning opportunities such as these, we argue that youth should be conducting community inquiry alongside people of a range of ages, diverse lived experiences, and different life trajectories and mobilities. This argument further highlights the inherent limitations of learning opportunities that only involve youth of similar ages or school grade levels.

Lastly, our third argument is that community inquiry as a learning opportunity should be framed within an asset perspective that acknowledges differences in the lives of youth not as deficits to overcome in comparison to an ideal cultural model, but rather as resources to be developed and complemented with additional resources. Community inquiry framed with an asset perspective "involves a reflection and investigation of what assets are validated and accepted, by whom, and under what contexts" (Ordoñez-Jasis & Jasis, 2011, p. 190). The asset perspective attunes community inquiry toward resources that may be overlooked or undervalued when considered through a deficit perspective. By focusing on community assets that are often overlooked, critical issues may arise that can challenge youth and other stakeholders to recognize the capacities of these formerly disregarded or undervalued community assets in relationship to the inquiry (Bruce, Bishop, & Budhathoki, 2014).

We draw on these three arguments to frame how we use mobile devices within community inquiry and to define what constitutes mobile learning. We consider the mobile device, in this case a GPS-enabled tablet, as a tool that can not only capture and represent information (e.g., pictures, video, sound, writing) that can be geolocated, but also a tool that can *travel* with the learner from place to place, while synching to virtual spaces. This physical and virtual mobility supports the distribution of learning within and across embodied experiences with people and places and in communication with remote people and spaces of learning.

Inquiry Cycle

Our pilot project followed an inquiry cycle developed by Bruce and his colleagues for community inquiry (Bruce & Bishop, 2002, 2008; Bruce & Davidson, 1996). The inquiry cycle includes five phases that are neither considered to be linear nor procedural: *ask*, *investigate*, *create*, *discuss*, and *reflect* (see Figure 6.1). In the following sections, we detail Adam's experience with each of these phases of the inquiry cycle. Here, we explain their general characteristics.

The *ask* phase begins with a question or problem that is related to the lived experience of the student conducting the inquiry. The *investigate*

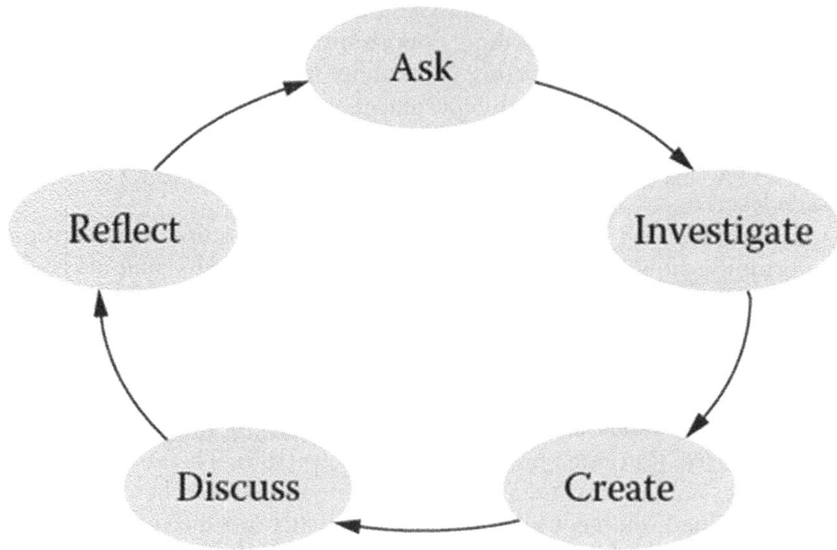

Figure 6.1. The community inquiry cycle.

phase presents opportunities for learning about the question or problem that may involve interactions with others and considering multiple data. The *create* phase involves the active creation of meaning, which may involve collaborating with others and taking on new roles to represent the learning with an artifact to be considered. The *discuss* phase involves engaging in dialogue with multiple stakeholders to consider how personal understandings relate to those of others. Lastly, the *reflect* phase involves taking stock of the experience and learning in order to outline future action. As we discovered in our work with Adam, the nonlinear characteristics of the inquiry cycle allow students to be simultaneously engaged in several of these phases at the same time.

COMMUNITY CONTEXT

The CIML pilot project was conducted in cooperation with a community service program for youth, Youth Engaged in Service (Y.E.S.). The Y.E.S. program serves participating adolescents ages 12-18 by providing volunteer opportunities to work with vulnerable members of the community (i.e., children, senior citizens, and the emotionally and physically challenged) in exchange for service hours that can be used for service-oriented organizations (e.g., KEY Club, Boy/Girl Scouts, National Honor

Society). Each of the five townships surrounding Buffalo, New York have a Y.E.S. Program, and our pilot project was conducted in one of these townships, West Seneca.

Due to delays in securing community approval for the pilot project, we began recruiting participants for the pilot study well after most families had made summer plans for their children. Adam was our only adolescent participant from the Y.E.S. program. As a result of this very limited enrollment in the pilot project, we changed the roles of some members of our research team from facilitators to participants. Ryan remained in the role of the pilot project facilitator. However, Abigail, Aijuan, and Merve, who are Ryan's graduate students, participated in the pilot project alongside Adam. We made this decision as not to overwhelm Adam with too many facilitators and to provide a social group with whom Adam could conduct his community inquiry. The five of us formed our research group around the shared goal of conducting community inquiry with the use of mobile tools. The group met for 10 sessions across three weeks at the West Seneca ice rink. Days that the group did not meet for a session served as opportunities for the group members to collect data using the mobile tools.

For the purpose of reporting on the pilot project, we decided not to anonymize the geographic location of Adam's community inquiry project in order to accurately represent the communities involved. However, we did decide to exclude geographic information within a half-mile radius of Adam's residence in order to protect his identity.

Adam's Mobility

At the time of the pilot project, Adam was an incoming senior in Orchard Park, a township south of West Seneca and without a Y.E.S. Program. Though Adam has lived in Orchard Park for over 15 years, visits to extended family and participation in ice hockey have extended his mobility into West Seneca on a regular basis. Adam participates in the West Seneca Y.E.S. Program for the purposes of serving that community and in part for the purpose of supporting his college applications. Throughout the community inquiry process, Adam acknowledged differences between Orchard Park, where he lives and attends school, and West Seneca, where he visits family and participates in sports not offered in Orchard Park, such as ice hockey. Some of these differences included socioeconomics and the types of opportunities that youth are able to take advantage of in each community. In Table 6.1, we provide the demographic data that Adam considered during his inquiry project (United States Census, 2010a, 2010b).

Table 6.1. Demographic Data

	West Seneca	*Orchard Park*
Population	44,711	29,613
Land Area in Square Miles	21.36	38.44
Population per Square Mile	2,093.5	755.8
Persons under 18 years	19.8%	23.7%
White	97.2%	96.7%
Black or African American	0.9%	0.7%
American Indian	0.2%	0.2%
Asian	0.6%	1.3%
Hispanic or Latino	1.7%	1.6%
Two or More Races	0.8%	0.8%
Median Household Income	$58,392	$81,326

Adam is representative of the majority of residents of Orchard Park, in that he is White and lives in a middle class household with his two parents. Though Adam primarily affiliated with Orchard Park as his home community, he regularly acknowledged that his experiences with family, sports, and service included the community of West Seneca. Therefore, Adam's mobility provided him with experiences that contrasted the socioeconomic differences between the two communities. Adam acknowledged his relative socioeconomic privilege in relationship to his friends in Orchard Park and his friends and extended family members in West Seneca (e.g., access to his own vehicle). Adam also acknowledged that his experiences across the two communities were not typical for the majority of his friends, who primarily remain in Orchard Park for their activities.

Therefore, Adam's experience challenged the initial design of our pilot study as involving adolescents from West Seneca who would conduct community inquiry within that community where we had established relationships with the Y.E.S. Program and the town board. Not only was Adam from Orchard Park, but he also had experiences that spanned both communities, which afforded him a contrastive perspective across socioeconomic differences among his friends. Though we originally conceptualized the pilot study based on participants' experience in the community in which they lived, Adam's experiences within and across West Seneca and Orchard Park expanded how we understood youth mobility and were defining community beyond municipal and school attendance zone boundaries.

Adam's physical mobility within and across the two communities presented an interesting opportunity from a mobile learning perspective.

Adam's daily travels put him in contact with people and learning opportunities that were not limited to municipal boundaries or school attendance zones. This mobility allowed Adam to connect diverse learning opportunities and people. The experiences also provided him with a contrastive perspective of what it meant to grow up and live within and across the two communities and a broader view of issues in the communities that could be investigated from his perspective for the project.

MOBILE ASSET MAPPING

During the overlapping *ask* and *investigate* phases of community inquiry, asset mapping involves the identification of community resources and learning opportunities, rather than initially focusing on what a community lacks or needs in comparison with some ideal (Kretzmann & McKnight, 1993). This approach intentionally counters deficit perspectives of people and communities by foregrounding overlooked and underappreciated assets that may escape the attention of stakeholders invested in bringing about change in a community. Example assets include, but are not limited to, people who serve as resources, facilities that offer opportunities, programs that offer services, and infrastructure that facilitates physical, virtual, and educational mobility. Once assets are identified, the community inquiry expands to consider who has access to these assets and how these assets can be complemented with additional resources and opportunities. This approach to community inquiry seeks not only to be responsive to existing community assets, but also seeks to sustain and not overextend them by simply making assets accessible to a greater number of people.

Mobile Tools

During the design of the pilot project, we considered a number of mobile applications that allow users to collect geographic, descriptive, and multimedia information without an active mobile data connection. In other words, we wanted to use GPS-enabled mobile devices (e.g., tablets, smart phones) without active data plans to collect information and then share that information through a wireless connection when available. This decision was motivated by our intent to scale the project eventually to include participants who may not have access to mobile devices with active data plans and to use older generation mobile devices that were donated to the project. We decided on the mobile application ArcGIS Collector because it allowed our participant to design a data form for collecting information and because it allowed our participant to present his data using ArcGIS Story Maps. In the following sections, we detail how Adam used Collector and Story Maps as we describe how his community inquiry unfolded during the pilot project.

ORIENTING ADAM TO THE MOBILE DEVICES

Adam came to the pilot project with a facility for working with mobile devices. He owns an iPhone that he uses for a variety of purpose related to his experiences with friends, family, and coaches in Orchard Park and West Seneca. Adam was familiar with using Google maps for navigation purposes. The GIS (global information systems) aspects of the pilot project were part of what convinced him to participate in the study at the encouragement of his parents. We provided Adam with an Android tablet for the project, which was GPS enabled, had wireless capability, and did not have an active mobile data plan. Adam easily adapted to the mobile device and operating system that was different than his iPhone.

In an effort to orient Adam to geolocation and basic GIS information, we asked him to use a geocaching application (i.e., c:geo) to locate and log three geocaches near the municipal park and ice rink in West Seneca where the pilot study was conducted. Geocaching is a recreational activity that involves using GIS information to find caches that are hidden by other participants. Caches range in size, but they typically consist of a container with contents that include objects of interest to the curators and a notepad or piece of paper to log the find within the cache. People engaged in geocaching who are using one of the online applications associated with a central database of the caches, can use their mobile device to log their find online and leave a note for the owner of the cache and other people participating within the central database. Adam was successful in his initial attempts to find the three caches we asked him to locate; though, his searching involved atypical physical mobility as he crossed a stream, cut through a baseball field, and traversed the park and neighborhood properties contrary to the ways the built environment was designed to be navigated.

We next asked Adam to create a cache that included items significant to him and to hide the cache in a place that was significant to his experience in the two communities. Adam selected a park near his home in Orchard Park. The cache Adam created represented aspects of his identity that were shaped by growing up in Orchard Park. In the small container that made up his cache, Adam included a golf ball, which represented his passion for sports, and a small metal gear, which represented his future plans to study engineering in college. These items in the cache are meant to be exchanged with other items by geocache participants who locate and log the cache. The location of the cache was also significant. Adam hid the cache in a park near his home where he played baseball as a youth and continues to go to socialize with his friends. Alongside Adam, the other members of our research group created and hid caches in places of significance. For the group, these activities related to geolocation and geocaching helped establish a relationship between the physical places we frequented and their geographic location on the representational mapping tools used in locating and logging caches. With Adam's cache in particular, we received feedback from other geocache players, who were competing to be the first to find (FTF) his cache. The players left comments in the online database that the GPS coordinates were not correct. As a group, we then returned to the cache and determined the correct GPS coordinates to record in the online database, which quickly lead to a player finding Adam's cache. We used these experiences with geocaching to learn how GIS tools can be used for geolocation and accurately representing assets on maps used in our inquiry projects.

ADAM'S INQUIRY PROJECT

Adam's inquiry project took shape based on his personal interests and conversations he had with our research team. As a research group, we took the approach of each discussing our interests in particular issues that are relevant to our respective communities. We did so by engaging in a map activity in which we drew and labeled maps by hand that included features that were significant to our experience and interests. Adam drew a map of Orchard Park which included streets with which he was familiar and omitted streets and neighborhoods with which he was not familiar. Adam's map also included labels of places that were significant to him, his family, and his friends. This hand drawn map, like the one he would later produce for his inquiry project, represents Adam's personal experience, mobility, and orientation to his community. We have come to call this type of geographic representation a "first person map," as it is not meant to serve as an objective representation of the community and is based on Adam's perspective and experience (see Figure 6.2). Once the maps were created, we discussed how we each represented our respective communities and why we chose to represent certain features over others. From these conversations, Adam made the decision to discover more about the opportunities for playing sports in Orchard Park and West Seneca based on his 15 years of experience growing up in the area. In the following subsections, we detail Adam's decision making and activities within each of the five phases of the inquiry cycle. However, Adam was often engaged in

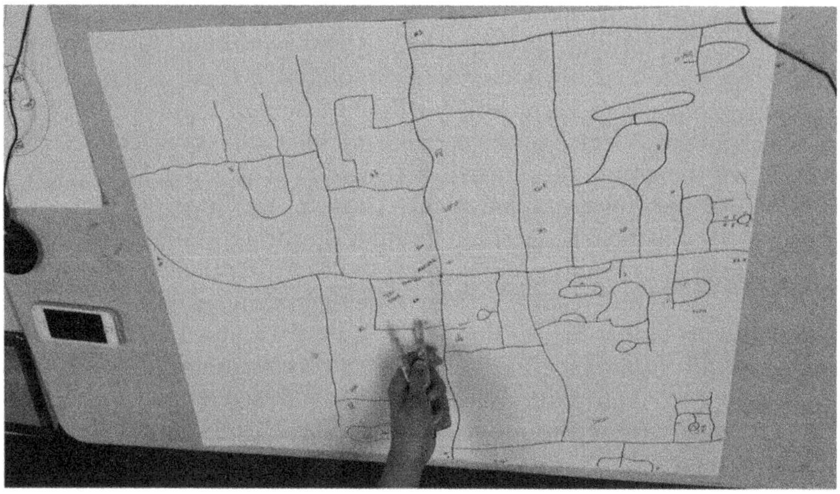

Figure 6.2. Adam's hand-drawn map of Orchard Park.

multiple phases simultaneously; therefore, we do not intend for our descriptions to be considered as procedural or linear.

Ask: Identifying the Sports Opportunities

Growing up in Orchard Park and West Seneca, Adam took advantage of playing a range of organized sports, including football, lacrosse, baseball, and ice hockey, as well as engaging in informal sport opportunities with his friends, including golf and tennis. Because our approach to community inquiry focused on assets at the onset, Adam set out to learn more about the opportunities for playing sports with which he was familiar and where these opportunities are located in proximity to where people live in Orchard Park. This first part of the inquiry cycle unfolded over two project sessions and represented an inquiry that arose from his personal experience (Bruce & Bishop, 2008).

We encouraged Adam to begin documenting his daily rounds to conduct an initial inquiry into the sports opportunities of which he was currently taking advantage. Adam was beginning his last year playing high school football as an incoming senior. At the time of the pilot project, he was beginning "two-a-days," or football practice that involved two sessions during the day. He was also working with a personal trainer in the northern village of Williamsville. Adam owns his own vehicle, which allows him to attend football practice, travel to his trainer and the pilot project site, as

Table 6.2. Daily Rounds Data Form Design

Fields	*Options*
Location Type	1. Exercise 2. Sports 3. Food 4. Leisure 5. Errands 6. Family 7. Work
Travel Type	1. Car 2. Bike 3. Walk
Description	N/A

well as meet up with his friends for recreation. In cooperation with Adam, we developed a data form for him to document his daily rounds using ArcGIS Collector. Adam defined the types of places he frequents and the ways he travels there. We built these categories into the data form with the addition of an open Description text field (see Table 6.2). These categories represent Adam's present mobility and access to opportunities within and outside his community. On a day our pilot project did not meet, Adam documented his daily rounds, which included three places that he documented using the Collector application. We instructed Adam not to include his family's residence, in order to keep his identity private when disseminating his work in the project. Figure 6.3 is a screenshot of the map Adam created with the data he collected during his rounds on that day.

The data from the first set of daily rounds helped Adam to develop a focused inquiry question that represented his interest in sports and his mobility. Two of the map points on Adam's daily rounds map were sports related: a map point for equipment check for the football season in Orchard Park and a map point in Williamsville, the location of Adam's personal trainer. Based on subsequent documentation of Adam's daily rounds and conversations our group had about Adam's experiences, Adam began to focus his inquiry on where sports opportunities for youth are located in relationship to where they live. However, the *ask* phase of Adam's inquiry cycle was not necessarily completed, as he would revisit the purpose of his inquiry during subsequent phases. In particular, when it came to the design of the data form for his data collection in the *investigate* phase, Adam revisited his central question in order to determine what characteristics of sports opportunities were salient for his inquiry.

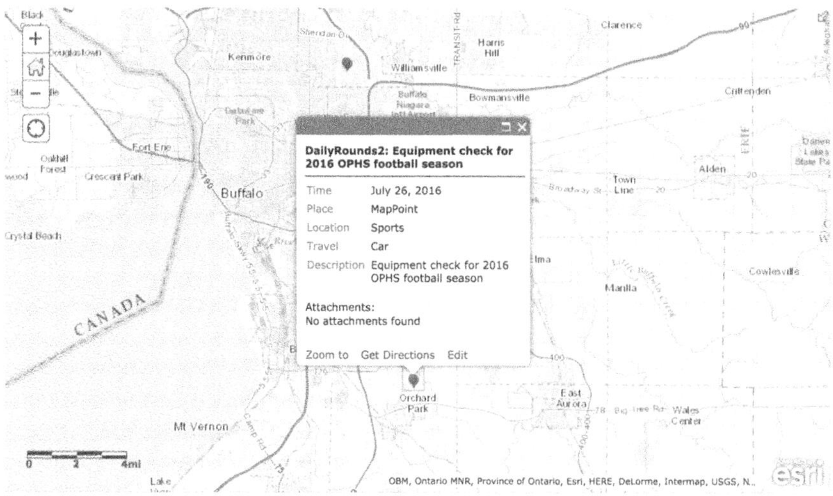

Figure 6.3. Map detail of Adam's daily rounds.

Investigate: Collecting Data About the Sports Opportunities

To support Adam's inquiry into what sports opportunities are available to youth in Orchard Park based on his experience, we discussed, as a group, what characteristics of these opportunities that he should document. We began making a list of all of the types of sports opportunities that Adam either took advantage of as a youth or was aware of through his friendships with other youth in the community. These places are represented in the first field in Table 6.3, which represents the design of the data form for Adam's use of the Collector app. We then discussed how many of these places involved play that was adult supervised (e.g., hockey practice at the ice rink) and how many were youth organized (e.g., wall ball at the school playground). We accounted for these differences in a field for type of play. Adam was also interested in whether or not a fee was involved in playing sports at these places. This interest developed out of a conversation about socioeconomic differences within and across the communities of Orchard Park and West Seneca. Lastly, Adam was interested in the accessibility of these places. Before Adam had access to a vehicle and a license to drive, he accessed some of these places by walking or riding his bike and accessed others by relying on vehicle transportation provided by his parents or friends.

We learned from our conversations with Adam that he had a more full and historical understanding of these places compared to what was repre-

Table 6.3. Sports Opportunities Data Form Design

Fields	*Options*
Place	1. Basketball Court 2. Baseball Diamond 3. Ice Rink 4. Football Field 5. Lacrosse Field 6. Soccer Park 7. Tennis Court 8. Track/Path 9. Golf Course 10. Pool 11. Gym 12. School 13. Other
Play	1. Adult Supervised 2. Youth Organized
Cost	1. No Cost 2. Fee
Accessibility	1. Walk or Bike 2. Vehicle
Description	N/A

sented by the maps. For example, when Adam was in middle school, he was limited to playing with friends whom he met up with in his neighborhood. Adam also explained that his decisions to go places were made, in part, by the distance from his home. This changed when Adam was given access to a phone and a vehicle, which afforded him both a way to arrange meeting up with friends and access to sports opportunities further from his home.

Additionally, some, but not all, of the sports opportunities that Adam documented are seasonal. For example, the outdoor facilities at the school can be used by all during the summer months, but use is limited when school is in session. Similarly, Adam's selection of the sports opportunities was shaped by the time of year, as he played hockey during winters in his youth, where he met friends who did not live in Orchard Park (see Table 6.3). These conversations and the design of the data form represented how Adam was coming to understand the complex factors involved in youth access to sports opportunities. However, the design of the data form presented some interesting issues in terms of representation as the options for the fields were mutually exclusive. The data form design did not allow Adam to capture some of the complexity involved with how these places were being used by youth in ways with which they were not designed or intended. For example, the data form did not allow

Adam to capture how the soccer field located at the elementary school was also the same place that kids went to play wall ball using the brick wall of the school. The baseball fields that were located within a municipal park also represented a location that was also repurposed as a hangout for Adam and his friends rather than a place to play baseball. Additionally, as Adam conducted his investigation he pointed out socialization opportunities that focused around the availability of food that were present at several of the places that featured sports opportunities. When we went on a walking tour of the sports opportunities that Adam was documenting, he made a point to show us where food was available in proximity to these places, indicating the significance of food when gathering to play sports or hang out with friends.

Adam's data collection with the Collector application both involved revisiting the sports opportunities he documented and placing pins on the map remotely, relying on his memory and the satellite imagery of the map layer he was using. These two ways of data collection provided related but different representations of the sports opportunities Adam documented. During our walking tour with Adam, where we revisited the sports opportunities he documented, we gleaned a fuller sense of what the place meant to him across time and how the place is situated geographically within the community. The guided tour served as an example of how youth embody place and build meaning in relationship to place through their movements and experiences over time.

Based on our walking tour with Adam, we came to understand that the sports opportunities Adam was documenting were as much about supporting sport as they were about supporting social relationships and an identity based on affiliations with particular places and sports opportunities within the community. Figure 6.4 depicts Adam taking a picture with his tablet that he attached to the pin he placed on the map using the Collector application. When we compare these conversations during the walking tour with the ones where we were looking at the representational maps at the pilot project site, we did not get as full a sense of what the place we were looking at using satellite imagery meant to Adam and the community. However, when we were on the walking tour listening to Adam, we did not have a good sense of how that particular place was related spatially and geographically to other places of significance to Adam. The spatial and geographic relationships were easier to understand using the representational maps. Therefore, we argue that both the embodied activity of visiting places to collect data and the use of representational tools, such as maps with satellite imagery, are of complementary importance during the *investigate* phase of the community inquiry. What we learned alongside Adam from these complementary sources of data speak to the value of collecting information remotely with mobile

Figure 6.4. Adam using the Collector app outside his high school.

tools and representing that information spatially and geographically on interactive maps.

Create: Making a Story Map

Across the days Adam was using the Collector app to document sports opportunities, we began discussing which demographic map layers and Story Map templates could contribute to Adam's inquiry. Adam was interested in where people lived in relationship to where these sports opportunities were present; therefore, we suggested that he consider population density as a possible layer to add to his map. Adam was also particularly interested in where youth lived, as well; therefore, we suggested that an additional layer could be population under 18 years of age. These suggestions were made in addition to others that he considered including socioeconomics represented by median household income, crime rate statistics, and public transportation services. Ultimately, Adam decided population density and population under 18 would best represent his interests to create his Story Map.

Next, Adam selected a Story Map template from those provided by ArcGIS for representing his experience. We discussed how each of these templates affords a different way of relating a narrative of a place and/or making an argument related to an issue that impacts a particular place. After considering the different options (see the Story Maps site at https://storymaps.arcgis.com/en/app-list/), Adam selected the Story Map Spyglass template because it allowed the user to reveal the population under 18 layer underneath the population density layer. Adam thought this would be the best way to compare the two map layers for a given neighborhood, in which his sports opportunity pins were placed.

Adam's last step in building his Story Map was writing the description that appears in the top left corner (see Figure 6.5). Adam wrote the description to orient his intended audience to the purpose of the Story Map and what the collected and assembled data represents. Adam understood his intended audience potentially to include municipal leaders, his friends who are interested in GIS and shared some of the documented experiences with him, and his family who were invested in particular places in the community (e.g., the youth football fields where Adam's dad coaches and his brother played). The building of the Story Map represents the active creation of meaning that Bruce and Bishop (2008) discuss as a key aspect of conducting community inquiry and the potential for communicating youth perspectives of place to community stakeholders. In particular, Adam was interested in demonstrating that not all of his friends may have had the same access he had to the sports opportunities he documented on his map due to where they lived.

Source: Adam's Story Map is available online at: http://arcg.is/2au3uEW

Figure 6.5. Adam's Story Map using the spyglass template.

Discuss: Shaping the Community Inquiry

The inquiry cycle is meant to be both based on the participants' lived experience and on dialogue with other people who are interested in the same or related issues. Bruce and Bishop (2008) describe the discussion phase of the inquiry cycle to involve "listening to others and articulating our own understandings" (p. 711). Because the inquiry cycle is not meant to be linear and the phases of the cycle may overlap, discussion took place on multiple occasions and in multiple settings during the CIML pilot project. This ongoing dialogue with the research group was essential to the meaning and decision making, as we worked together to shape and focus the inquiry and make decisions about the use of the mobile tools and the design of the data forms for data collection.

At the onset of the pilot project when we were making our hand-drawn maps, we discussed decisions we were making in representing particular issues in our respective communities. In this initial part of the project, Adam drew from our dialogue to begin making choices for conducting his own community inquiry. For example, our early discussions involved a consideration of modes of transportation and physical mobility in our respective communities. Our research group represented urban, suburban, and rural areas, and we compared differences in physical mobility and transportation used to navigate our respective communities. Adam drew directly from these conversations to include physical mobility and transportation both in the documentation of his daily rounds and the accessibility of the sports opportunities that he represented in his Story Map.

These discussions about mobility and transportation also helped us understand that Adam's community inquiry was not bound by assumed boundaries based on municipal borders and school attendance zones. We had assumed that his inquiry would be focused on a single community, Orchard Park, but through our discussion we came to understand that Adam's history with sports opportunities involved mobility and transportation outside of Orchard Park to the neighboring townships of West Seneca and East Aurora.

Subsequent discussions during and after Adam's building of his Story Map revealed that Adam was making connections between his lived experience and the data he was collecting with the Collector application and assembling with the Story Map template. Adam shared that growing up he had a sense for where youth lived in relationship to where opportunities for sports were situated; he explained that this spatial orientation shaped where they played, with whom they played, and the places where they coordinated their play. Adam said that the population density and population under 18 data confirmed and nuanced what he understood to be true about his community, in terms of where youth lived in relationship to where opportunities for sports were situated. Depending on where youth lived in Orchard Park, opportunities to play sports and coordinate activities with other youth varied. These mobilities shaped relationships for Adam and his friends that began in their elementary school years and continues to shape relationships in their high school years. For example, Adam's closest teammates on the football team came from the same area of Orchard Park and played together on the same youth football team.

Reflect: Considering the Significance

At the end of the pilot project, we had an opportunity to reflect as a group on the significance of Adam's community inquiry. We emphasized that this round of inquiry could inform subsequent inquiry for Adam or someone else in the two communities with related concerns about where opportunities for sports are situated in proximity to where youth live. As such, we reflected on what Adam had accomplished with his inquiry and the building of his Story Map as much as we did about the different directions the inquiry could take in the future.

We considered the significance of using the tablet as a mobile tool to collect data about the sports opportunities by visiting those places. During our walking tour with Adam, he helped us understand that visiting those places to collect the data both allowed him to consider features that he may not have remembered, but also allowed him to make connections to other considerations such as his familial and social relationships that took shape on those fields. The embodied activity of collecting mobile data was as significant as using the representational tools to create the maps.

Adam currently plays high school football with friends with whom he played as a youth, and his younger brother now plays on the same fields on which Adam played, coached by their father. The football fields took on a greater significance in Adam's community inquiry as he reflected on how it had been a hub of activity for his family and friends over the years during football season. Though the design of the data form for data collection using the Collector app did not allow Adam to capture this history and complexity, his visit to the football fields and the taking of pictures during our walking tour did facilitate these considerations of how sports opportunities shape familial and social relationships in his community. We discussed how to redesign the data form to collect temporal data (with the use of time stamps) in addition to spatial data (with the use of geolocation) in order to document change over time with the Story Map templates.

When we asked Adam who he thought would be interested in his Story Map, he mentioned a friend who was interested in political maps and also his teachers who may not be familiar with the community in which they teach. Adam explained that he noticed a difference between his teachers who lived in the community and had a history there compared to his teachers who commuted to his school from other communities. Adam considered a teacher's knowledge of and history within the community to be an asset, because he felt better understood by those teachers. Adam considered that his Map Story may be a way for teachers unfamiliar with the community to learn more about how sports opportunities shape social relationships and help define the community through his personal experience. Adam also speculated that teachers who are unfamiliar with the community would benefit from taking a walking tour with their students, like the ones we took with Adam.

Bruce and Bishop (2008) explain that "reflection means expressing experience and thereby being able to move from new concepts into action" (p. 711). We understood this phase of the inquiry to also include what actions Adam could take in the future if given the opportunity to further his community inquiry beyond the limits of the pilot project. Adam shared that he was interested in the relationship between authority figures (e.g., police, business owners) and youth both inside his community and in other communities with which he was less familiar. Adam reflected on his own experience with authority figures in his community, such as getting chased out of restaurants for being loud or being asked to relocate to another area when playing, and wondered about the relationship between authority figures and youth in other communities where there have been unfortunate incidents of police officers killing unarmed youth. At the end of the pilot project, Adam was expressing interest in issues experienced by youth in communities that are different from his own. Adam came to see the community inquiry project as a viable way for youth to provide spatial representations of perspectives and experiences to community stakeholders that they may not otherwise have access to.

DISCUSSION

Our research team approached this project keeping in mind that it was a pilot and that, as a group, we were fellow learners in the learning opportunities we were supporting for Adam. We were interested in how mobile asset mapping could support community inquiry conducted by youth. Together, we learned a great deal about the significance of the embodied activity related to collecting mobile data and the use of representational mapping tools to represent issues spatially in the community. Below, we revisit our three arguments about learning opportunities that informed our approach to engaging Adam in community inquiry: connecting multiple places of learning, orienting toward the lived experiences of participants, and framing within an asset perspective.

Adam's learning processes related to the community inquiry project not only took shape in specific places (e.g., pilot project room in the ice rink, places in the community), but his learning processes also took shape along trajectories and physical, virtual, and educational mobilities that connected those places. During the three weeks of the pilot project, Adam collected data out in the community on days when he was not meeting with us, and he discussed with us the significance of the data he collected on days when we met as a group. The inquiry took shape not only in the specific places where Adam collected and discussed his data, but the inquiry also took shape across them as we developed new questions together and revised the data form Adam was using to collect data. We feel that it is significant that Adam's learning was not only informed by the places he went to collect data and discuss the project, but also by new understandings and considerations as he moved from place to place. For example, collecting data in one place about what sports activities are supported there (e.g., baseball fields) led to new considerations of other places (e.g., brick wall in the school yard) in terms of what sports were played there, despite the intentional design of the built environment of that place.

Adam's community inquiry began with his lived experiences growing up playing sports in Orchard Park and West Seneca. Adam was aware that his experiences were not the same as other youth in those communities, and he was interested in where the opportunities to play sports were situated in relationship to where people lived. As a research team, in the planning stages of our pilot project we were interested in critical social issues related to race and socioeconomics, in particular, issues related to the legacy of segregation in Western New York. We were a bit unsure when Adam decided to focus on sports opportunities, but we trusted the design of our pilot project and followed Adam's lead to focus on an issue relevant to his life. We discovered that his inquiry did ask critical questions about access to opportunities and how those opportunities shape social relationships in his community. Adam was quick to point out instances where his experience between the two communities of Orchard Park and West Seneca was different than his friends who primarily participated in activities within Orchard Park. Adam's community inquiry work with the mobile and mapping tools provided him with spatial data to support his understanding of how neighborhoods and school attendance zones shape which youth play sports together and which sports opportunities they have access to. Adam considered his contrastive perspective within and across the two communities to provide him with a vantage from which he considered issues of access and opportunity in his community shaped by socioeconomics. Our research team learned much we did not know about the two communities through Adam's inquiry project. This helped us in turn consider our own relationship to the two communities, as we develop the CIML project beyond the pilot phase in partnership with the schools and communities of West Seneca and Buffalo.

Adam's inquiry focused on community assets, in terms of what opportunities for sports were provided by the community. This focus on assets helped Adam account for places for sports opportunities that were not part of his experience. Though he may not have played a particular sport or used a particular facility, he collected data about these places in his inquiry as relevant to his inquiry question. These places were known to him through his friendships and mobility in the community. Adam's asset-based data collection also revealed the ways that youth use facilities in ways they were not designed nor intended to be used. For example, the brick wall in the school yard may be overlooked by community stakeholders as a significant opportunity for sports, but Adam's perspective as a former youth who played wall ball there revealed the place as a significant asset as a place for sport and as a place for social gathering. These considerations ultimately led Adam to consider the relationship between youth and authority figures, which he described as a possible next step in his inquiry project for helping the community understand the perspective of youth who want to gather and organize sporting opportunities.

In sum, Adam's community inquiry represented a learning opportunity for the research group that was distributed within and across multiple places, involved mobility within and across these places, and was shaped by many different people with whom Adam communicated. We view Adam's experience as a salient case of mobile learning, as he used the mobile tools and mapping software to conduct his inquiry and represent his findings. Most significantly, the community inquiry project supported Adam in learning a new way of inquiring into personal experience and critical issues that relate to community development. Our research group is also indebted to Adam for helping us think through how to support community inquiry with mobile asset mapping as we scale the project to include one of the high schools in West Seneca.

IMPLICATIONS

We view Adam's community inquiry and Story Map less as a model or mentor text and more as a representation of what is possible with combining aspects of community inquiry and mobile learning. Had we more time with Adam, we were interested in helping him develop his project further to include a range of issues and series of maps that collectively would represent more of the complexity of his experience growing up in Orchard Park and West Seneca. In particular, we were interested in drawing on map layers that represent socioeconomics at the neighborhood level to further support Adam's observations that sports opportunities were often not located in the most densely youth populated areas. We were also interested in involving more adolescents to contribute to the mobile data collection and build a collective and contrastive spatial representation of youth perspectives on opportunities for sports within and across the two communities.

As we scale this project to involve more adolescents and purposefully partner with classroom teachers and community stakeholders, we draw on Adam's experience with the community inquiry project about sports opportunities. Adam's experience challenged our definitions of community as bound by municipal borders and school attendance zones. His experience also reminds us of the importance of beginning the inquiry cycle with personal interests and issues that directly impact the youth with whom we are working. Lastly, we are reminded that Adam's access to a vehicle and freedom his parents allowed afforded him a physical mobility that most likely will not be the same for all youth with whom we plan to work. These considerations challenge us to frame these types of learning opportunities in ways that are appropriate for the youth, schools, and communities with whom we plan to work.

REFERENCES

Bruce, B. C., & Bishop, A. P. (2002). Using the web to support inquiry-based literacy development. *Journal of Adolescent and Adult Literacy, 45*(8), 706–714.

Bruce, B., & Bishop, A. (2008). New literacies and community inquiry. In J. Coiro, M. Knobel, C. Lankshear & D. Leu (Eds.), *The handbook of research in new literacies* (pp. 699–742). New York, NY: Routledge.

Bruce, B. C., Bishop, A. P., & Budhathoki, N. R. (Eds.). (2014). *Youth community inquiry: New media for community and personal growth*. New York, NY: Peter Lang.

Bruce, B. C., & Davidson, J. (1996). An inquiry model for literacy across the curriculum. *Journal of Curriculum Studies, 28*(3), 281–300.

Kretzmann, J. P., & McKight, J. L. (1993). *Building communities from the inside out: A path toward finding and mobilizing a community's assets*. Chicago, IL: ACTA.

Leander, K. M., Phillips, N. C., & Taylor, K. H. (2010). The changing social spaces of learning: Mapping new mobilities. *Review of Research in Education, 34*(1), 329–394.

Ordoñez-Jasis, R., & Jasis, P. (2011). Mapping literacy, mapping lives: Teachers exploring the sociopolitical context of literacy and learning. *Multicultural Perspectives, 13*(4), 189–196.

United States Census. (2010a). Welcome to QuickFacts: Orchard Park town, Erie County, New York. Retrieved from https://www.census.gov/quickfacts/table/INC110214/3602955277/accessible

United States Census. (2010b). Welcome to QuickFacts: West Seneca town, Erie County, New York. Retrieved from http://www.census.gov/quickfacts/table/PST045213/3602980918/accessible

CHAPTER 7

WE GOT THIS

Toward a Facilitator-Youth "Apprenticeship" Approach to Supporting Collaboration and Design Challenges in Youth-Designed Mobile Location-Based Games

Sara Vogel
The Graduate Center of the City University of New York

Judy Perry
Massachusetts Institute of Technology

It's May at a New York City high school's after-school program. We—the adult facilitators—have been guiding a group of youth to produce a location-based mobile game. The teens have worked dozens of hours and want to see a finished product. But with the final playtest coming soon, we are stuck. Reflecting on the most recent session with the participants, we realize that the game is far behind where it should be. Despite having already begun to code the game and write its interactive text, the game's core mechanics are still only half-baked, and adjusting problematic real-world locations in the digital game will take time we just do not have. Something went wrong in the design process, and we have to figure out what to do next.

Mobile Learning: Perspectives on Practice and Policy
pp. 143–168

We ask ourselves, should we make this a "teachable moment?" Reiterate the challenges of mobile game design and let them experience what happens when your product does not work and you are facing a deadline? Or should we adults step in, potentially undercutting the teens' agency as designers, to help them achieve a playable outcome? Time is ticking. We decide that this time, we'll step in and mock up a new prototype to help them out. But there has to be another way to balance the challenges of designing a complex mobile game against the teens' agency and ownership over the process. What might we do better next time?

With the release of *Pokémon Go!* in 2016, geolocative or location-based augmented reality (LBAR) gameplay experiences have become far more widespread and familiar. Mobile LBAR games embed players in an experience, providing layers of digital information (characters, objects, and interactions) displayed on the player's location-aware smartphone based on her current location. As players move around the real-world game space, virtual game components enable dynamic, meaningful experiences with the attributes, artifacts, landscape, and cultures of the physical spaces (Klopfer & Squire, 2007). Due to the mobile nature of LBAR games, their creators put their own stamp on a place—from a local urban neighborhood to a curated museum space—constructing within and commenting upon real-world contexts in their own distinct voices (Klopfer & Sheldon, 2010). As designing for mobile devices becomes more accessible, youth are becoming empowered as producers rather than merely consumers of this technological genre.

In this chapter, we describe a study in which researchers from the Massachusetts Institute of Technology (including Perry) and practitioners from the New York City-based youth development organization Global Kids Inc. (including Vogel) worked together on the iterative development of an out-of-school-time program in which high school-aged young people engaged in a collaborative design process to craft mobile LBAR games. While many elements (setting, duration, scope, resources, and age group, among other things) contribute to the outcomes of such a youth program, this chapter focuses on the critical role *adult facilitators* play in structuring activities and supporting youth to overcome the particular challenges of collaborative design of mobile LBAR games.

The adult facilitators of this program followed a design-based research methodology (Anderson & Shattuck, 2012), which engages researchers and practitioners in systematically designing, reflecting upon, and iterating the implementation of an educational program or intervention in an authentic educational context. Following this methodology, not only did we refine our curriculum design, but also our theories about the role of adult facilitation in such projects. With promotion of youth ownership and youth voice as core values undergirding our program, our initial impulse as adult facilitators was to only lightly scaffold activities and intervene infrequently. However, as a result of iterative design, we as practitioners came to recognize that our initial assumptions about structure and agency were overly simplistic. Rather than viewing structure and agency in opposition to one another, we modified both our thinking and our practice similarly to the way Brennan (2013) argues that appropriate structures enhance learner agency. Specifically, to ensure youth were engaged in productive collaboration around meaningful design challenges, we as adult facilitators needed to—at key moments—become active collaborators working in partnership with youth following what Kirshner (2008) would call an apprenticeship model. This is consistent with conclusions reached in other youth development contexts (Bolstand, 2011; Larson, Walker, & Pierce, 2005; Kirshner, 2008) where objectives for youth require them to operate just outside of their Vygotskian zone of proximal development (Vygotsky & Cole, 1978). We found this apprenticeship approach to be especially necessary in our context, given the complex, multilayered nature of the mobile artifact we were expecting youth to produce.

WHAT SKILLS CAN LBAR GAME DESIGNERS GAIN?

For more than 2 decades, researchers including Kafai (1994) have argued persuasively that game design offers youth opportunities to gain valuable insights and experiences as they grapple with meaningful challenges throughout the design cycle: planning, prototyping, testing, and iterating upon a game concept. If game producers code their own games, they also gain computational thinking and computer programming skills (Wang & Chen, 2010). Mobile technology specifically affords opportunities for games that are place-based, which means youth designers of these games also become more familiar with their environments, learning geospatial skills and geolocative contextualization (Klopfer & Sheldon, 2010). Designers also stand to gain storytelling and graphic design competencies. The more aspects of the LBAR game design process that youth are exposed to, the greater the potential benefit.

At an organization like Global Kids, which values and promotes youth expression, youth produce place-based mobile games that reflect their local experiences and histories, and transmit messages not often seen in mainstream games or media. When participants work together to author these games, they also learn to collaborate. LBAR game creation is an example of a complex design task which affords students authentic opportunities to move from being autonomous problem-solvers toward working with others to achieve a common goal (Jenkins, Purushotma, Clinton, & Robison, 2009). The collaboration fostered through LBAR game design might also support character development (Coulter & Stauder, 2015).

WHAT CHALLENGES DO LBAR GAME MAKERS FACE?

We expected participants to develop a range of competencies and skills from collaboratively designing a mobile LBAR game. But that would not be without its challenges.

Challenges Stemming From the Novelty and Complexity of the LBAR Design Process

Despite being avid players of many kinds of games, youth participants of Global Kids' programs generally had no experiences with LBAR mobile games (Pokémon Go! had not yet been developed at the time of this study). Facilitators would be responsible for introducing youth to the genre. In addition, facilitators expected that the design process would be new to many of our participants. While some youth regularly engage in peer and interest-driven design practices as they "mess around" and "geek out" with digital technologies (Ito et al., 2008), in traditional school-based settings, youth typically encounter few opportunities to produce complex products with new technologies in an open-ended way, and, thus, may struggle with the open-ended nature of designing place-based AR (augmented reality) experiences (Mathews, 2013). Students may even resist such constructivist approaches to learning to create with digital tools because of the disjuncture with traditional school expectations and cultures (Brennan, 2015). Despite being roughly organized around a process that cycles through ideation, prototyping, building, and testing, design processes are fluid, nonlinear and unpredictable (Lawson, 2006).

LBAR games, in particular, are multilayered products; especially those designed to promote learning. They often integrate several instructional approaches, including emphasizing locations, the player's role, and the tasks she must complete (Wu, Lee, Chang & Liang, 2013). The AR experiences that Global Kids encouraged youth participants to design fell into a category that Klopfer (2008) calls "lightly augmented," in which players interact with specific features in the physical environment. Therefore, during the ideation phase, youth designers of lightly augmented mobile experiences engage in place-based education to make thoughtful connections to particular locations (for an extensive literature review on place-based education in youth-produced AR contexts, see Mathews, 2013).

Once LBAR designers gain familiarity with a particular location, they draft and geolocate an appropriate narrative with meaningful connections to that place and produce an effective prototype of the game. The creation of prototypes is a particularly challenging, important step of the game design process for youth (Macklin & Sharp, 2012). A good prototype helps designers test, and therefore ensure, that the parts of the system (interactions with place, story, game mechanics, and theme) are coordinated to provide an optimal experience (Mathews, 2013).

Challenges Arising From the Collaborative Nature of the Project

A second set of challenges arose from the collaborative nature of the project. Youth AR experience designers interviewed in Mathews' (2013) study ranked conflicts around consensus-building as some of the most difficult aspects of the design process. To operate efficiently and democratically, members of collaborative groups must be skillful at "providing everyone with a chance to speak, coordinating the actions of group members, reaching consensus, ensuring elaboration of the material being learned, and keeping all members on task" (Johnson & Others, 1984). Collaboration is a skill learned over time, and in classroom contexts, teachers have roles to play in scaffolding group work (Gillies & Boyle, 2010). Our goals around collaboration included ensuring all of the teen participants at our sites felt like they contributed to and had a voice in the project. We also aimed to guide the youth to recognize that in collaborative design, compromise is inevitable and conflict can be productive.

OUR CONTEXT

The settings chosen for study were summer or after-school programs run by Global Kids, Inc. Global Kids is a non-profit youth development organization that provides out-of-school time programming, in-school enrichment, and other services "to ensure that youth from underserved areas have the knowledge, skills, experiences and values they need to succeed in school, participate effectively in the democratic process, and achieve leadership in their communities and on the global stage" ("Global Kids | Home," n.d.). Most staff are professionals in the field of youth development, many from the communities they serve. In aiming to foster leadership skills, the organization urges staff to promote youth self-expression, be sensitive to the diverse needs of participants, and ensure workshops are inclusionary and youth-led whenever possible.

In 2011, Global Kids and the New York Public Library founded NYC Haunts, a STEM-based learning program in which youth designers create a digital, mobile location-based game exploring local history and contemporary issues facing a particular neighborhood. The NYC Haunts location-based games produced by Global Kids youth typically follow the footsteps of a "ghost"—a historical figure or composite modern-day character with some "unfinished business" for the player to assist (Vogel, 2014). The games aimed to immerse players in stories that would truly engage with the history, local issues, or artifacts in a place. To create their games, youth used TaleBlazer, a software platform developed by the Massachusetts Institute of Technology's Scheller Teacher Education Program Lab. The software is equipped with a blocks-based programming language and a user-friendly interface—features that enable young people and other non-experts to code the back end of these games (Medlock-Walton, 2012).

Research was conducted during three Global Kids' NYC Haunts programs during the spring and summer of 2014 (program components and participant demographics summarized in Table 7.1), which all served approximately the same number of high school-aged youth (between 14 and 16 youth participants). Of the 45 total participants, 20% of them reported their ethnicity as Latino/a, 67% of them Black, 7% Asian and 7% White. Fifty-seven percent of them were female, and 43% were male. Youth were engaged in the programs for roughly the same number of contact hours (between 24 and 25 hours). The spring program (Iteration 1 of the study) differed from the summer programs (Iteration 2 of the study) in a few key ways. Participants in the spring semester-long afterschool program received partial school credits. Attendance was sometimes spotty. Participants in the month-long summer programs, which were run in conjunction with an art museum in Brooklyn and a community center in the Bronx, received a stipend (museum) or minimum wage pay (community center). Attendance was generally consistent in the summer.

Several adult facilitators helped to shape this study's design and implementation (see Table 7.1). Across the three sites, there were a total of six adult facilitators. The spring program was led by two educators and the summer programs were each led by four. Vogel, one of the authors of this study, was the lead educator facilitating the program across all sites and programs. In the spring, she was joined by a Global Kids facilitator who had worked with some of the youth in previous programs. At each summer program, Vogel was joined by one facilitator and a college-aged intern from Global Kids, and one of two site-specific facilitators. It was the first time that facilitators worked with students at the museum. The community center-based staff member had prior relationships with some of the youth there. Facilitators varied in years of experience working with

Table 7.1. Summary Description of Youth Game Design Program Sites Included in the Study

Iteration	*Site*	*Schedule*	*Attendance*	*Demographics*	*Facilitators*
1	High school (Brooklyn)	Jan.-May 2014, weekly 1.5 hr. sessions after-school, ~25 contact hours	16 enrolled, 11 attended a majority of sessions, 7 consistent attendees per session, unpaid; partial credits in school	10 girls, 6 boys (1 Latino, 15 Black)	2 total: Vogel and one Global Kids on-site educator
2	Site A: Community center (Bronx)	Jul.-Aug. 2014, biweekly 2.5-3 hr. sessions during a summer work program, ~25 contact hours	15 enrolled, 13 consistent attendees, compensation (minimum wage)	9 girls, 6 boys (8 Latino, 7 Black)	4 total: Vogel, one Global Kids facilitator, one college intern and one community center educator
2	Site B: Art museum (Brooklyn)	Jul.-Aug. 2014, biweekly 3 hr. sessions, ~24 contact hours	14 enrolled, 14 consistent attendees, compensation (small stipend)	7 girls, 7 boys (3 Asian, 8 Black, 3 White)	4 total: Vogel, one Global Kids facilitator, one college intern, and one museum educator

youth and varied familiarity with the TaleBlazer technology and the game design process. All facilitators ascribed to the philosophy of promoting youth voice. All three sites benefited from at least one visit from Perry, who supported as a consultant.

CONDUCTING OUR RESEARCH

In line with design-based research approaches, as we designed our original curriculum and considered its implementation and outcomes, we reflected on our baseline assumptions about the role of the facilitator. There are many potential models for adult-youth interaction in youth programs, with some approaches classified as more "adult-driven" and others more "youth-driven" (Larson et al., 2005). Approaches along this spectrum have their benefits and limitations. Youth-driven approaches might be more appropriate for programs that expect youth to take an active role in decision-making and that are geared toward the development of leadership skills, whereas adult-driven approaches often afford more opportunities for youth to develop specialized skills or organize sophisticated events or projects (Larson et al., 2005). Larson (2006) outlines the tensions facing adult facilitators of youth programs: adult participation in the process should not be so heavy-handed or directed as to limit or stifle youth motivation, engagement, voice or ownership over the project. At the same time, too little guidance from mentors can leave youth participants feeling without direction or purpose, or result in incomplete projects that youth are not likely to take pride in.

In some youth-digital media production settings, such as the Computer Clubhouse centers around the world, youth are encouraged to use software and other materials to make projects of their choosing and adult mentors are "guides on the side," engaging in little direct instruction of concepts. In these contexts, "the creation of a design culture requires substantial support and direction from Clubhouse coordinators and mentors, less so in telling Clubhouse members what to do but more so in helping them develop their own ideas" (Kafai, Peppler, & Chapman, 2009, p. 4). In other contexts, such as in video production, adults have managed tight schedules (Goodman, 2003), provided critical feedback (Goldman & Booker, 2008), and have "polished" work so that outputs are high-quality products (Jenson, Dahya, & Fisher, 2014). In Macklin and Sharp's (2012) examination of "issues literacy"-based game design, facilitators of one program chose to limit the parameters of the project to focus the youth primarily on research and producing a working paper prototype, rather than the back-end coding and development: "We need to consider balance, one of the most important hallmarks of a challenging and fun game, as we try to balance the triad of issues: literacy, game design, and technology/tools" (Macklin & Sharp, 2012, p. 400). Echoing these approaches, before beginning our first iteration, our intention was to provide some structures for youth participation; but in line with our philosophies about youth empowerment at the time, we aimed to do so minimally and infrequently.

Using the design-based research technique of conjecture mapping (for more detail see Sandoval, 2014), we developed a conjecture to both articulate and evaluate design and theory, and later reshaped the conjecture as our research unfolded. Initially, during Iteration 1, our conjecture was that *facilitators providing limited structure would allow for a maximum of equitable youth active participation (agency) and learning through engaging with the full design arc of constructing a location-based augmented reality game.* Our curricular facilitator materials and practices (described in more detail in the next section) embodied this conjecture by encouraging facilitators to provide limited structure for student efforts in keeping with our level of experience with AR game creation at this time. However, during Iteration 1, facilitators saw that students were often overwhelmed by the degree of novelty and complexity of the design process, which meant facilitators had to vacillate from a "hands off" approach to a high-involvement, "save the day" approach doing the heavy-lifting when the students were appreciably floundering. As described in the next section, we saw the need for additional structure in two main areas: (1) student collaboration and (2) scaffolding the design process. Despite the increased facilitator involvement in Iteration 2, we hoped the project would remain largely youth-driven.

After completing the Iteration 1 of the program, we were struck by the degree to which we hoped for, but did *not* see, youth collaboratively grappling with a range of meaningful design challenges unique to mobile LBAR game production, such as closely observing and integrating features of the landscape into the game, taking active roles in crafting the game's story, translating paper prototypes to code in TaleBlazer, and problem-solving around coding bugs. We also observed that a few dominant voices (including ours as facilitators) had an outsized role in the end product, even though we had hoped that a range of voices about the place and its history would be synthesized to create a broader whole. In line with our conjecture, we had initially so lightly structured design activities that the situation necessitated a drastic ramping up of scaffolding at the end in order to yield an experience that approached our initial vision of engaging youth in the entire arc of the design process.

We considered where students were struggling and determined that not only did we need to change aspects of the curriculum itself, but also our underlying assumptions about our roles as facilitators. In debriefing our experience, we found a model to help guide us in formulating a new approach: Kirshner's (2008) research on adult-youth interaction during programs centered around activist campaigns. Kirshner conceptualized three categories of adult-youth interactions: *facilitation*, *apprenticeship*, and *joint work*. In the facilitation approach, adults sought to be neutral, "hands-off" guides of a youth-led process. The apprenticeship approach was characterized by adults engaging in campaigns alongside youth, while modeling, scaffolding, and structuring complex activities in ways that were sensitive to youth skill levels. The *joint work* approach is a form of collaboration where adults and youth work together on a project in an environment where the goal is oriented more toward completing the project than on training novices. We began to view our approach to supporting youth LBAR game creation through the lens of **apprenticeship**. While Kirshner found that in programs exhibiting this approach, youth had fewer opportunities to plan and facilitate meetings than in approaches where adults were less involved, participants in the program which tended toward apprenticeship had "more extensive practice developing and implementing a campaign with clear policy objectives" and "gained access to a communication strategy used by seasoned advocacy organizations" while still benefiting from a youth-centered environment (p. 84).

Revising our conjecture, we recognized that rather than providing limited structure, *facilitators might "apprentice" the youth by providing more consistent, thoughtful structures, still very much aimed at maximum equitable agency and learning through engagement with the design process of LBAR games*. We refined our practices at various steps, providing more structured support for collaboration. We also engaged in codesigning alongside the youth, specifically, in structuring and supporting work at the early stages of developing a LBAR game (e.g., brainstorming, synthesizing ideas, and prototyping). As we discovered, these first stages are crucial moments which can set youth up to succeed or stall as they embark on producing a particularly complex, multilayered artifact. Our specific scaffolding and structuring practices will be discussed in depth in the next section below.

Throughout the implementations, a variety of data were collected to shed light on adult-youth interaction, collaboration between youth members of the group, the opportunities they had to grapple with design challenges, and their sense of ownership and voice over the process and final product. Data collection methods included pre-, mid-, and postsurveys of youth participants, postprogram interviews with youth participants, an analysis of products created by the youth, and facilitator fieldnotes taken after each program session. A sampling of youth participants also participated in semistructured interviews by Vogel, Perry, and Jason Haas, a doctoral student at MIT. Finally, photographs of the various products of the design process were analyzed, including brainstorming sheets, anchor charts, game design documents, prototypes, slide decks created for the final presentation, and the TaleBlazer games themselves.

IMPROVING PROGRAM DESIGN ACROSS ITERATIONS

In both the spring and the summer at all three program sites, youth emerged from the design process with a playable LBAR game artifact that leveraged mobile technologies to link gameplay to local environments and culturally relevant topics. In the spring, the game took place at a housing complex near the youth participants' high school, which, decades prior, was the site of Ebbets Field, home of the Brooklyn Dodgers baseball team. In this game, the player assumed the role of Jackie Robinson, who broke the color barrier in baseball in the late 1940s at that stadium. Youth participants found the task of creating a mobile game in their community eye-opening, with one participant commenting, "most people that live in

Figure 7.1. Screenshots of the introduction (left) and a nonplayer character (NPC) (right) in "Paths of the Forgotten: Ghosts of Hunts Point." This location-based mobile game was produced by youth as part of a summer program at a community center in the Bronx, NY.

the apartment complex ... they don't know about Ebbets Field ... or how important is was.... When I think back ... it's really disappointing that I lived in a place for so long and I didn't know anything about it." Youth at the museum-based summer program produced a spooky game which took place at the cool, dimly lit visible storage exhibit. Here, the player is an average museum visitor who must, through close observation and selection of particular museum artifacts, uncover how a fictional girl named Helen died in order to halt her haunting of the exhibit. At the community center's program, participants designed a game (Figure 7.1) centered around Drake Park, a recently discovered slave burial ground from the 17th and 18th centuries in the middle of Hunts Point in the Bronx. The player must restore items to the "forgotten" ghosts of the slaves in order to honor their memories and use location-based clues to guide a runaway slave to safety.

In postprogram interviews and surveys, youth from all three sites expressed that they were proud of the game they produced and their contributions to the final outcome. All 12 of the youth interviewed from both spring and summer programs reported that collaboration and teamwork were some of the central takeaways of the program. The kinds of challenges they described having grappled with throughout the design process, however, were quite different.

Several key moments in the spring (Iteration 1) design process challenged our ideas about the optimal role for facilitators of mobile LBAR game design programs. Below, we describe structures that facilitators put in place at various moments during the spring program's game design process and report the visible practices of the youth participants connected to those structures. We then describe modifications to these facilitation structures for the summer programs, and the youth practices that we observed subsequently.

Brainstorming, Evaluating, and Selecting Ideas for the Game

Coming to consensus around choosing a game's topic, location, and story-premise was a challenging endeavor for many youth participants in both the spring and the summer. As Rachel,[1] a 17-year-old female participant at the museum, put it:

> I'm not really the collaborative type. Like, I normally like to do things on my own and get them done because I feel I work well when I know what I want to do. I can just get right to it. So, it took me a while to adjust to the collaborating with others and hearing their ideas and stuff.... It took me a while to realize I couldn't do this on my own.

Anticipating that the youth would face challenges around generating, selecting, and evaluating ideas for their game, facilitators put specific initial structures in place to scaffold this work in the spring later modifying these practices for the summer programs.

Iteration 1—Spring Approach: Unwittingly Fostering Competition

In the spring, facilitators attempted to guide students to brainstorm and select a topic for their game as a group. The facilitator's practices included (1) encouraging youth to advocate for their own particular, disparate ideas and (2) facilitating open-ended whole group discussions.

The first practice, youth advocating for their disparate ideas, may have unwittingly fostered a competitively-oriented dynamic among the participants. After a whole-group information-share and brainstorm about local events and history, youth chose to do preliminary research about distinct and diverse topics. Some had chosen to look into Revolutionary War history in the area, others were interested in crime history, and another was passionate about Jackie Robinson's role in baseball and Black history in the area. Facilitators encouraged the youth to go around in a circle, sharing their ideas with the group. They were told that if they felt strongly about one or another of the topics, they should make a case for it by justifying their ideas using a set of criteria that the facilitators had generated and posted on the board, which included ensuring the game was about local history, took place around the school's neighborhood, was structured around a story, and required the player to walk around, meet characters, answer questions, and/or collect items. Ideas were shared, but one voice emerged as dominant. Fatima, a 14-year-old female participant, made a case for her idea related to Ebbets Field and Jackie Robinson, using a combination of criteria she had developed herself (related to the cultural relevance of the topic) and criteria from the facilitators (related to local history and location). In describing that moment, Fatima interpreted the process as a competition:

> They all had spectacular, I mean really great ideas—I mean really great—ideas and we all wanted our ideas to be the one chosen. Of course, that is what a competition is, but after I talked to them for a while, they realized that we should do something much more close, and that was my main argument: something much more close and something that relates to all of us. And most of the other topics, they related to us, but not as directly as the topic of Ebbets Field and Jackie Robinson.

At the end of the game design process and program, the group's final presentation highlighted Fatima as the participant responsible for the game's topic. When two other students were interviewed about the topic selection process, they both attributed the idea for the game to Fatima. Lawrence, a 15-year old male, said that Fatima's idea "overshadowed every other topic." While other youth game designers had bought into the idea, they continued to view it as "her" idea, up to the end. By facilitating a "voting" model across widely varying topics, there could only be one clear "winner" rather than a thoughtful synthesis, and ultimately few contributions or voices were incorporated.

Having selected a topic for the game, facilitators engaged in the second practice: open-ended whole group discussion. As the youth worked toward the basic story, goal, and mechanics for the game, adult facilitators asked the whole group to answer questions (e.g., *What should the player's goal be? What does the player do in the game?)* to spur on an organic conversation about ideas. The first such group discussion was fruitful, "with students contributing ideas like baseball card collections, radio announcements and other details" (Field notes, 3/26/2014). However, discussion at the next design session became unwieldy, as captured in our field notes:

> To some extent they were building on each other's ideas, but sometimes it was as if they hadn't heard the idea before theirs and contributed unrelated ideas. There was a range of ideas expressed: finding something Jackie Robinson had lost, finding a genie in a piece of wood, collect cards because someone is stuck in purgatory. it [sic.] was difficult to evaluate all of them and slow students down...The dialogue was free-flowing, but at times, criticism was bordering on unsafe space territory. Some students were not open to criticism. Other students usually quite active (Shamar) today during the fast-paced design process seemed withdrawn toward the end. (Field notes, 4/2/2014)

The loose, unstructured format facilitators chose for discussion made it difficult for the students to tease out, evaluate and incorporate ideas meaningfully. While we had intended to promote youth agency with structures that would provide space for participants to advocate for their ideas in a whole group, the end result was an atmosphere where a few voices dominated, while others retreated. All signs led to reevaluating facilitator practices toward more inclusive and participatory collaboration during this hurdle in the design process.

Iteration 2—Summer Approach: Structuring Compromises

To promote a wider range of youth voices and ownership over the topics, locations, and stories of their games, we changed the structures for generating ideas and selecting a topic during the summer programs. The three main practices we employed were (1) choosing locations and game topics all together, (2) encouraging groups of youth designers to brainstorm around similar locations and topics, and (3) modeling how ideas could be separated from their initial origins, evaluated, and recombined.

First, instead of splitting designers up to brainstorm about disparate topics and locations, game locations were established together as a whole group before dividing up to brainstorm. We hoped that, in that way, youth would brainstorm story ideas around more similar themes, thus making it possible to combine multiple ideas into a cohesive synthesis, rather than having to choose one and discard others. At the museum, the group had a chance to tour three different exhibits and to select one to house their game. Small groups were randomly assigned to record observations and evaluate the pros and cons of each exhibit as a potential game space. Youth evaluated spaces using a set of criteria which the facilitators had modified from the spring program to be more open-ended and reflective

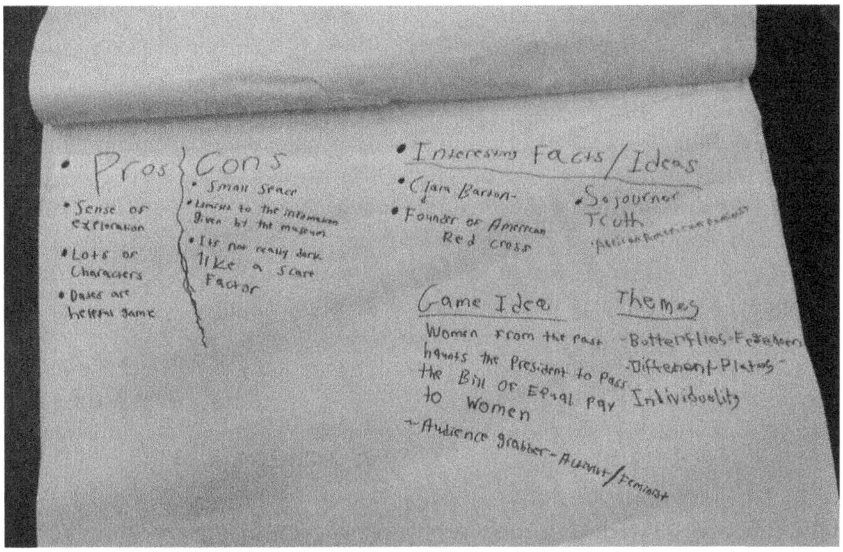

Figure 7.2. Youth in a summer implementation evaluate the pros/cons of a potential game location during a group brainstorm at the museum. Ideas are voiced, included, and synthesized from a range of participants.

of strong location-based games. These criteria were also elicited from the youth by facilitators after they had played a sample location-based game, and included judging whether place was used in a unique way (i.e., could this game be played anywhere else?), whether the gameplay was fun, and whether the story supported gameplay. At the same time, the process of touring and evaluating the exhibits also helped the youth develop new criteria to evaluate potential spots, stating that one location promoted a "sense of exploration," and had "lots of characters" (Figure 7.2). Participants still had an opportunity to justify their perspectives about the best locations before taking a final vote, but those youths whose ideas were not selected were still able to contribute core ideas to the game because the only element we had determined up to that point was the game's location.

At the community center, the location was preestablished by the staff there to be Drake Park, the former slave burial ground. We worried that this choice would undercut youth agency and ownership over the game. In midway and postprogram surveys, when asked to answer the question, "How interested are you in making a game about this topic?" by indicating their response on a Likert scale of 1 (*not at all*) to 5 (*a lot*) four of the participants who answered the question ranked the topic with a 4, and three with a 3. Only two indicated 5, "a lot." Contrast that with the responses from the museum program, where all but one participant marked "5," and the remaining marked a "4." Some at the community center did not much mind that the topic was preselected, as one 18-year old participant, Sofia, noted "I'm a person who just goes with things, so I was like, 'Okay, Drake Park. I'm fine with it.'" Another participant in the 11th grade, Janice, however, noted on her survey, "We were given the topic by facilitators/not really given a chance. Not that history isn't interesting but I was never so into social studies." Given these responses, we would have to exercise caution in this regard in the future. Not only did it undercut some participants' ownership over the game, but unlike the museum participants, they did not have the opportunity to practice evaluating locations against criteria. The youth generally rated other aspects of the program highly, however, and choosing the topic in advance did ensure that youth would have the time to make collective decisions about the game's story and mechanics.

To facilitate brainstorming and topic selection, we implemented a second practice: convening small groups of youth designers to generate potential story ideas and game mechanics appropriate to that location. Facilitators purposefully kept the brainstorming parameters loose to solicit a range of ideas. Designers were told to think of ideas for a game which would fit well within, and teach players something about, the selected location. Youth were permitted to draw, write, and talk about their ideas. The goal was not to develop polished, internally coherent ideas, but to generate a variety of ideas. Figure 7.3 illustrates one group's collectively brainstormed ideas for the game near the community center.

Each small group then presented their ideas to the larger group. As they did so, facilitators engaged in the third core practice: modeling for youth how ideas could be separated from their initial origins, evaluated, and recombined. We jotted notes down on poster papers, separating out ideas that referred to the player's role, the player's goal, the agents (virtual characters) the player would meet, and game mechanics. After each small group's presentation, other participants were given an opportunity to state which components of the group's ideas they liked, and to explain why. Ideas that were "seconded" were checked off on the poster paper (Figure 7.4). After the presentations, individual youth designers were given an opportunity to state their "must-haves"—the ideas they felt most passionate about.

This approach to evaluating ideas communicated a crucial aspect of the design process: the fact that ideas generated during brainstorms might be recombined in new ways. This key theme around synthesis of ideas was reflected in the participant interviews during the summer program. Eighteen-year-old Ron, a male participant at the community center, said his favorite part of the project was "everybody interacting and

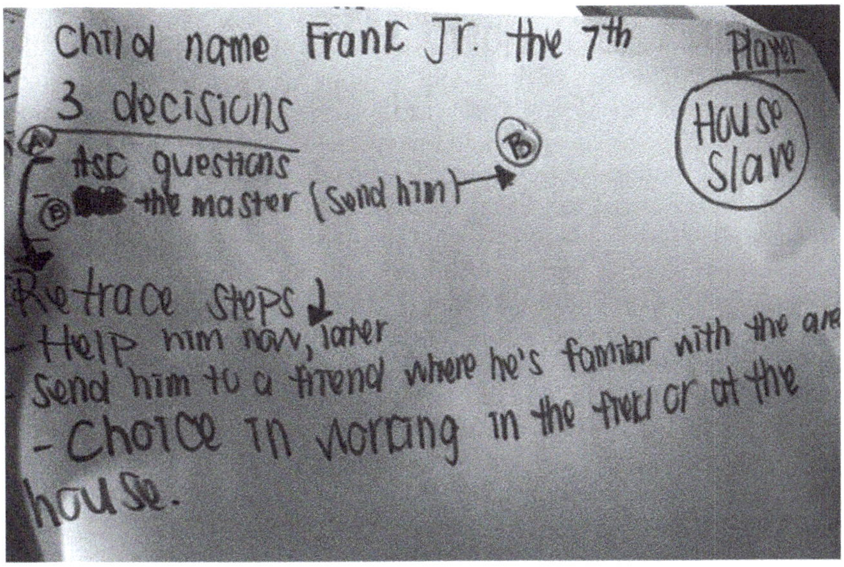

Figure 7.3. One of many pages of ideas generated by youth during a summer implementation at a community center. Participants brainstormed numerous preliminary ideas for a potential mobile game, enabling a wide number of voices to contribute ideas.

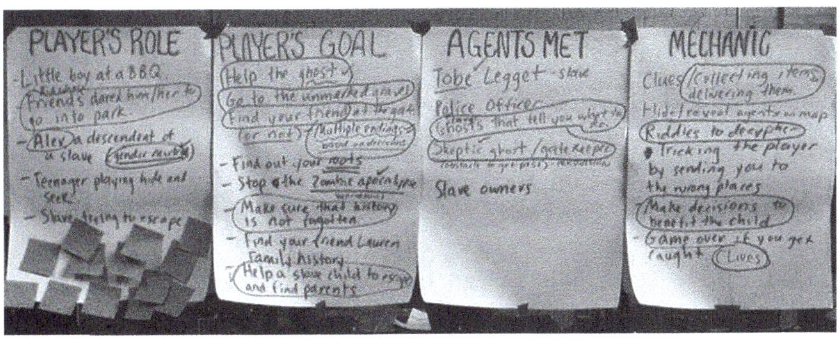

Figure 7.4. Facilitators organized youth ideas into relevant categories and created structures for youth to "vote" for components of a mobile game design which they liked, allowing ideas from multiple youth designers to potentially combine into a single game.

everybody ideas come together to make one idea." In his interview, Trevor, a 17-year-old male participant at the community center characterized the process: "we just kept on throwing out ideas and eventually all the ideas merged into the game we have today."

Participants still took ownership over the ideas they contributed. Sofia, 18, of the community center program, was proud that her group "got picked to have the basis of the game" adding, "we put a lot of time into it and they picked our idea. That was really awesome." Even if one group's general premise was chosen, there was space for others to also feel proud of their contributions. For example, Trevor said: "Um, I don't wanna take the credit for it, but I feel like, I brought up the idea of having multiple scenarios." A few youth interviewees also took pride in watching their ideas evolve. Zora, a participant at the museum program said, [I like] "having my ideas put into something. Like, if I say something, people try to work with it." At the same time, she said that she became more open to other participants' ideas throughout the course of the program, noticing that "when my idea wasn't picked and someone else's was, I realized it was better than mine." When asked what collaboration in this project looked like for her, Rachel, of the museum program, said: "Definitely compromise. Like, collaboration for me is, like, everyone sort of comes with their ideas and we all talk about our ideas and then it's just a matter of, 'I like this and your idea and my idea and let's use these two things and make a whole new idea.'"

Our approach in the summer was to evaluate and select locations together, encourage youth designers to brainstorm around similar locations, and then to model how ideas could be separated from their initial origins and recombined. These structures made it easier to synthesize ideas—providing more opportunities for youth to see their own contributions reflected in the game premise. Such an approach also prompted more opportunities to watch ideas evolve and improve.

Designing the First Paper Prototype to Test the Game Concept

In the early stages of the design, our participants faced yet another key challenge: moving from a smattering of ideas to a viable paper prototype that successfully integrated TaleBlazer-friendly game mechanics, location, and narrative. During the first iteration in the spring (as described in the vignette at the start of this chapter), youth as well as adults found ourselves flummoxed and stalled at this stage. By the summer, armed with more experience and context, we could support youth in tackling this challenge earlier, fast tracking them over a key hurdle and enabling them to grapple with challenges related to building the digital game.

Iteration 1—Spring Approach: A Shaky Foundation

At the high school site, the youths' game premise about Jackie Robinson breaking the color barrier at Brooklyn's Ebbets Field had many elements of a strong game: a main character and auxiliary characters, conflict, and roots in both the neighborhood location and in history. They had a few ideas for mechanics including answering trivia questions in order to collect "supporters," and stumbling upon hidden agents that would trap players or provide a bonus. As the weeks went by, facilitators came to recognize the significant holes in the game premise: What would these trivia questions be about? Could we assume the average player could answer these trivia questions? Would trivia questions at all engage the location where the game would be taking place? Facilitators articulated these questions to youth, and worked with them to brainstorm an alternative: "ethical dilemmas" rather than trivia questions. Youth then began to work on coding and writing the story for that premise, discovering the following (as noted during a facilitator check-in):

> As the "research/story" team came up with the character descriptions and dialogues, they kept pointing out how obvious the answer to the [ethical] question (did I make the right decision?) actually was. After students wrote their stories using powerful language about Jackie Robinson's struggles, the questions even sounded condescending. Students and adults alike came to the conclusion that the design of the game / the questions had to change. (Field note, 5/7/2014)

In debriefing what had occurred, we realized that the game's concept had been weak from the start, but both facilitators and youth were unsure how to strengthen it. We attempted to guide the youth to design a paper prototype for their project, but the one we came up with was incomplete, failing to adequately incorporate the question types as well as critical aspects of the game's location. In previous workshops guiding the construction of TaleBlazer games, games were less complex, relying less upon prototypes to uncover and iterate through problematic designs. However, in the spring groups, with youth taking the lead in creating far more complex games, facilitators felt that creating a paper prototype might be beneficial. However, with limited experience making and using prototypes, we didn't scaffold youth to create robust enough prototypes capable of surfacing deficiencies within the designs. As such, we moved forward with the game idea before evaluating whether the idea successfully integrated all parts of a location-based game (location, narrative, mechanics, theme). By the time the facilitators (including Vogel) sought help from mentors at Global Kids and TaleBlazer (including Perry) to produce a prototype that captured the essential elements of the game concept, we were just weeks from the launch and had to dramatically compress the production timeline. Facilitators rushed the youth to complete the game's story, coding, and artwork, providing very explicit templates for code and game text for the students to work off of, leaving less time and space for youth to express their creativity and to grapple with unique challenges of the building stage.

The compressed design timeline did enable the group to finish the program with a playable prototype for their game, which they took pride in playing and showing to others. However, rushing at the end also limited youth designers' experience with other meaningful design challenges including grappling with storytelling, coding, and outdoor playtesting. In the spring, all four youth participants who included design challenges on their midprogram and postprogram surveys mentioned one thing: reconciling ideas among the group / choosing topics. In interviews, three of the four youth spring participants recalled challenges the group faced in the paper prototyping stages. They expressed that their initial ideas for the game were too complex, that the questions they wanted to ask were too obvious, and recalled how we attempted to come up with solutions, but they attributed the ultimate solution to Judy (Perry) and the adult facilitators, rather than to their own efforts. During interviews, youth at the spring after-school site remembered being part of a storytelling or coding team, but did *not* cite specific challenges.

Iteration 2—Summer Approach: Fast-Tracking Youth Over the Synthesis Hurdle, Transparently

In the summer, the approach to synthesizing the group's ideas involved more hands-on work from the facilitators *at the outset,* rather than as a scramble at the end. It also involved being transparent with the youth about the work the facilitators did. After the facilitators took stock of the ideas that the youth generated and flagged during their brainstorms, we decided to meet together, apart from the participants, to synthesize their ideas into a playable paper prototype that would integrate place, story, and game mechanics, from the start. Our goal as facilitators was to incorporate as many of the youth designers' "must have" ideas as possible—adding few to no new ideas so that the youth would see their own contributions reflected back—while also ensuring coherence and clarity of game elements.

Youth mostly needed assistance in selecting a TaleBlazer-friendly core mechanic that would serve as an organizing principle for the game. For example, at the community center, participants had suggested that ghosts of the slaves who were in unmarked graves at the site should be characters in the game, and that potential goals of the game could be to "help the ghosts," "make sure that history is not forgotten," and "save your friend." They brainstormed mechanics such as "collecting and delivering items." Armed with the youth's must-haves, facilitators were able to map out a prototype to structure their ideas. In order to connect participants' ideas about not forgetting this history to the "collect/deliver" mechanic they had suggested, we incorporated a "memory" mechanic. Players would meet a series of characters, all of which would be ghosts of the slaves buried around the park. They would learn what items each ghost desired to honor their memories. The player would then have to remember which item each ghost needed, delivering the correct items to them.

At both sites, facilitators synthesized youth ideas during after-hours. When the groups reconvened the next day, facilitators presented youth with a visual outline/paper prototype to express the core elements of the game (Figure 7.5). We discussed transparently which ideas from the brainstorming session we had incorporated and modeled our thinking about how we integrated multiple, interrelated components. Youth were encouraged to ask questions about, evaluate or change the prototype as we presented it. At the museum, facilitators presented the group with two paper prototypes so youth designers might choose whether the game would have a linear or nonlinear structure. Youth in both locations incorporated the facilitators' synthesis of ideas as a point of departure for fleshing out particular storyline elements and mechanics.

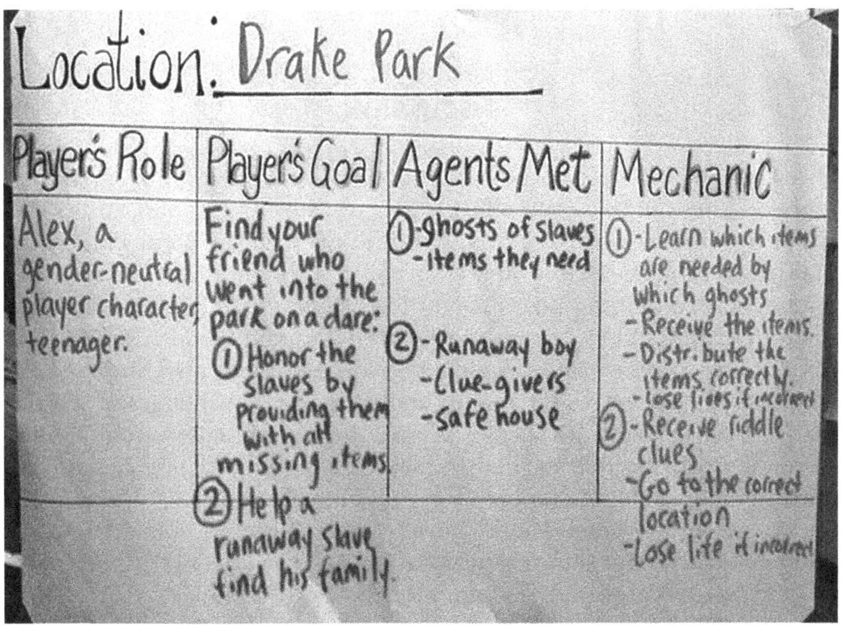

Figure 7.5. A proposal for a mobile game structure consisting of youth's ideas, as synthesized by facilitators.

This fast track over the "synthesis hurdle" also afforded youth participants more time to focus on actually building the game. In the summer, interviewees and survey respondents wrote about a greater range of meaningful design challenges such as using a flowchart to guide their logic as they coded, debugging scripts, and making collective decisions about whether to include controversial subject matter. There was also more time for youth to engage with the locations where the games would take place. A smaller team within the larger group was able to visit the sites where their games would take place multiple times, closely observing elements in the space, and working within the constraints of the location to select features in the landscape that would best tell their stories. As a result, both mobile games produced during the summer programs integrated place in more intentional ways than the spring game did.

While youth participants had played a sample game and had been exposed to the TaleBlazer game editor, they were still relative novices in this domain. The facilitators—having learned from experiences in the spring—had more knowledge about the range of possible game mechanics that the editor would support, and we were able to lend our expertise to get youth over a key hurdle in the process. The working prototypes that facilitators brought to the table helped youth make informed choices and made it possible for the youth to skip some of the structuring steps so they would have time for other aspects of the design process.

DISCUSSION

Given youths' close connection with mobile technologies, the creation of mobile games can be a compelling way to transform technology users into technology makers, and to promote youth's deep engagement with particular places. Mobile devices' limited graphics capabilities, smaller screens, and modest computing power might lead those unfamiliar with the process of creating mobile games (particularly location-based AR games) to initially view this task as a simple one. However, these assumptions belie the inherent challenges in creating mobile games, challenges which our program participants discovered. These false expectations weren't limited to youth game creators. Facilitators, new to designing mobile games, similarly initially failed to appreciate the difficulty of this challenge. It became clear, over time, that the novelty and complexity of making LBAR games required us to shift our perspectives and to view ourselves as learners and collaborators in the design process.

We can consider varying levels of facilitator support for youth LBAR game creation along a continuum (Figure 7.6), with lower levels of support (left) building toward higher levels of support (right). During the spring

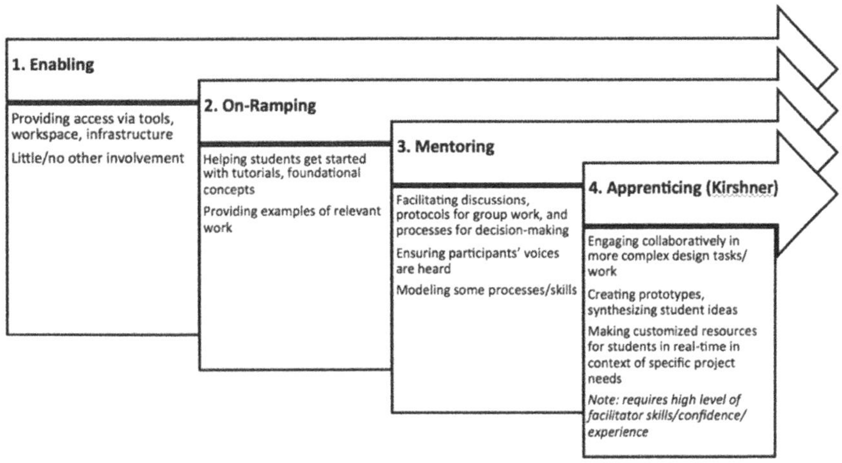

Figure 7.6. A model of increasing levels of potential adult/facilitator support, from least (left) to most (right) supporting collaborative youth creation of LBAR games.

implementation of the program, initial expectations were that we might focus on levels 1–3 of the continuum: enabling youth to build by providing resources, on-ramping them with examples, and then mentoring them with structures for group discussion and collaboration. We were hesitant to go beyond this, lest we tread too heavily on youth agency. However, it became clear that rather than expecting youth participants to effectively navigate these complex tasks, facilitators instead might change their overall approach, turning to more of a level 4 apprenticeship model (Kirshner, 2008). In assuming this perspective, we facilitators not only gave ourselves permission, but saw value in stepping in to support tasks above the current level of participants, modeling skills and helping to develop strategies and structures to collaboratively find solutions to roadblocks.

Some youth participants we interviewed recognized the value of a "we got this" approach as well. As Trevor, a summer participant from the community center, put it:

> I feel as though, it's [the game is] all of our souls. I guess you could say that. Like, the instructors and ours, because we did a lot of work, but they weren't like, 'You guys got it.' Everyone put a lot of work in.... It was just collaborative the whole time...Where they were at, I like it because they would push me to find the answer on my own.

Once facilitators internalized their roles within this apprenticeship model and let go of the false notion that all decisions ought to be guided single-mindedly by the north star of youth agency, we were free to consider ways to ensure youth maintained ownership over their projects and the process. Even as we framed ourselves as collaborating guides, we still had to take caution not to overstep our role as adults outside of the collective from which "the work" would emerge, as evidenced by some of the community centers' youth participants' reactions when the location and topic were chosen for them. It helped when we were transparent about our own contributions and privileged youth ideas and voices over our own.

Ultimately, this study shifted our practice as facilitators of mobile game creation toward a more supportive approach from the spring to the summer implementations. This support was possible because the adult facilitators meaningfully scaffolded the design process, enabling the group to collaborate more effectively and experience a fuller arc of the game design process. Yet, while facilitators provided more support, they did so within an apprenticeship model with the goal of developing youth capacity to eventually take on complex collaborative tasks independently. Ultimately, these efforts empowered youth to make deeper connections not only with each other, but also, through the medium of a mobile location-based game, with their local communities.

ACKNOWLEDGMENTS

Special thanks to the stewards of the NYC Haunts program, Barry Joseph, Juan Rubio, and Jack Martin, to the adult facilitators and youth participants of the program, and to the executive director of Global Kids, Evie Hantzopolous. Thanks to Jason Haas for assistance with interviews, Alora Cholette and Andres Pardo for support with transcriptions, and Glenda Stump, Meredith Thompson, Ling Hsiao, and Rafi Santo for their feedback at key moments in the drafting of this project.

NOTE

1. All names of youth are pseudonyms.

REFERENCES

Anderson, T., & Shattuck, J. (2012). Design-based research: A decade of progress in education research? *Educational Researcher, 41*(1), 16–25.

Bolstand, R. (2011). From "student voice" to "youth-adult partnership." *Research Information for Teachers, 1*, 31.

Brennan, K. (2013). *Best of both worlds: issues of structure and agency in computational creation, in and out of school* (Doctoral thesis). Massachusetts Institute of Technology, Cambridge, MA.

Brennan, K. (2015). Beyond right or wrong: Challenges of including creative design activities in the classroom. *Journal of Technology and Teacher Education, 23*(3), 279–299.

Coulter, B., & Stauder, R. (2015). To have fun and display our awesomeness: mobile game design and the meaning of life. In C. Holden, S. Dikkers, J. Martin, & B. Litts (Eds.), *Mobile media learning* (pp. 139–144). Pittsburgh, PA: ETC Press.

Gillies, R. M., & Boyle, M. (2010). Teachers' reflections on cooperative learning: Issues of implementation. *Teaching and Teacher Education, 26*(4), 933–940.

Global Kids | Home. (n.d.). Retrieved from http://globalkids.org/?view=mobile

Goldman, S., & Booker, A. (2008). Mixing the digital, social, and cultural: Learning, identity, and agency in youth participation. In D. Buckingham (Ed.), *Youth, identity, and digital media* (pp. 185–206). Cambridge, MA: MIT Press.

Goodman, S. (2003). *Teaching youth media: A critical guide to literacy, video production & social change*. New York, NY: Teachers College Press.

Ito, M., Horst, H., Bittanti, M., Boyd, D., Herr-Stephenson, B., Lange, P. G., … Robinson, L. (2008). *Living and learning with new media: Summary of findings from the digital youth project*. Retrieved from http://eric.ed.gov/?id=ED536072

Jenkins, H., Purushotma, W., Clinton, K., & Robison, A. (2009). *Confronting the challenges of participatory culture*. Cambridge, MA: MIT Press.

Jenson, J., Dahya, N., & Fisher, S. (2014). Valuing production values: A "do it yourself" media production club. *Learning, Media and Technology, 39*(2), 215–228.

Johnson, D. W., & Others, A. (1984). *Circles of learning: Cooperation in the classroom.* Alexandria, VA: Association for Supervision and Curriculum Development.

Kafai, Y. (1994). *Minds in play: Computer game design as a context for children's learning*. Hillsdale, NJ: Routledge.

Kafai, Y. B., Peppler, K. A., & Chapman, R. N. (2009). *The computer clubhouse: Constructionism and creativity in youth communities*. New York, NY: Teachers College Press.

Kirshner, B. (2008). Guided participation in three youth activism organizations: Facilitation, apprenticeship, and joint work. *Journal of the Learning Sciences, 17*(1), 60–101.

Klopfer, E. (2008). *Augmented learning: Research and design of mobile educational games*. Cambridge, MA: MIT Press.

Klopfer, E., & Sheldon, J. (2010). Augmenting your own reality: Student authoring of science-based augmented reality games. *New Directions for Youth Development, 128*, 85–94.

Klopfer, E., & Squire, K. (2007). Environmental detectives—The development of an augmented reality platform for environmental simulations. *Educational Technology Research and Development, 56*(2), 203–228.

Larson, R. (2006). Positive youth development, willful adolescents, and mentoring. *Journal of Community Psychology, 34*(6), 677–689.

Larson, R., Walker, K., & Pearce, N. (2005). A comparison of youth-driven and adult-driven youth programs: Balancing inputs from youth and adults. *Journal of Community Psychology, 33*(1), 57–74.

Lawson, B. (2006). *How designers think: The design process demystified*. Thousand Oaks, CA: Routledge.

Macklin, C., & Sharp, J. (2012). "Freakin' hard": Game design and issue literacy. In C. Steinkuehler, K. D. Squire, & S. Barab (Eds.), *Games, learning, and society: Learning and meaning in the digital age* (pp. 381–402). Cambridge, UK: Cambridge University Press.

Mathews, J. M. (2013). *Place-based design: An instructional design theory for supporting community-based inquiry and design projects* (Doctoral thesis). The University of Wisconsin-Madison, Wisconsin.

Medlock-Walton, M. P. (2012). *TaleBlazer: A platform for creating multiplayer location based games* (Thesis). Massachusetts Institute of Technology, Cambridge, MA.

Sandoval, W. (2014). Conjecture mapping: An approach to systematic educational design research. *Journal of the Learning Sciences, 23*(1), 18–36.

Vogel, S. (2014, January 16). NYC Haunts, now at a GK school near you! | Online Leadership Program. Retrieved from http://olpglobalkids.org/content/nyc-haunts-now-gk-school-near-you

Vygotsky, L. S., & Cole, M. (1978). *Mind in society*. Cambridge, MA: Harvard University Press.

Wang, L.-C., & Chen, M.-P. (2010). Learning programming concepts through game design: A PCT perspective. In 2010 Third IEEE International Conference on Digital Game and Intelligent Toy Enhanced Learning (DIGITEL) (pp. 219–221)

Wu, H.-K., Lee, S. W.-Y., Chang, H.-Y., & Liang, J.-C. (2013). Current status, opportunities and challenges of augmented reality in education. *Computers & Education, 62*, 41–49.

CHAPTER 8

AUGMENTING NATIONAL HISTORICAL PARKS

A Pilot Study for Harpers Ferry

Laura A. Gillespie
University of Baltimore

"I John Brown am now quite *certain* that the crimes of this *guilty land*: *will* never be purged *away*; but with Blood. I had *as I now think*: *vainly* flattered myself that without *very much* bloodshed it might be done."

— From John Brown's last letter,
the day he was hanged, December 2, 1859

Exploring Harpers Ferry National Historical Park in West Virginia can be a profound experience as one observes the natural beauty of the Blue Ridge Mountains and the Potomac and Shenandoah rivers, the picturesque remains of a 19th century industrial town, and the vestiges—some barely visible—of a federal government presence that included one of only two U.S. armories at that time (National Park Service [NPS], 2009). Outdoor exhibits situated around the town distill the various stories of the area down to a few sentences. These are fascinating stories to be sure, but

Mobile Learning: Perspectives on Practice and Policy
pp. 169–187

remote and somewhat decontextualized. Visitors can stand among the historical remains and consider the events described, but how can they grasp the enormity of that town's significance to abolitionist John Brown and the impact of his ill-fated raid in 1859—a raid that famed orator and abolitionist Frederick Douglass refused to endorse, and the possible success of which he completely dismissed (Frye, 2009; NPS, 2009; Varon, 2008)? How can visitors understand, let alone empathize with, Brown's decision to go through with it anyway? Finally, how can they understand this raid as one of the galvanizing events leading into the American Civil War (Frye, 2009; NPS, 2009; Public Broadcasting Service [PBS], 2000; Varon, 2008)?

This is perhaps a tall order considering the raid occurred over 157 years ago. What remains of that time is a fairly robust material culture, including primary sources, artifacts, numerous structures, and the general topography of the town of Harpers Ferry (itself significant); however, enough has changed in those 157 years that National Park Service ranger and Harpers Ferry's chief historian Dennis Frye remarked that, were John Brown to come back to Harpers Ferry today, he would hardly recognize the place (D. Frye, personal communication, April 2013). We are therefore already at a disadvantage in trying to accurately reconstruct this particular past; however, learning about the past and taking away valuable lessons from it does not necessarily require pinpoint accuracy. Educators can help learners perceive, in a broader sense, the events of the past through the exploration of a physical historical site, and through role play, step into the opinions, struggles, and decisions faced by others from another time and consider what they might have done in their place. Augmented reality (AR) and mobile games can help make that connection.

The intersection of mobile devices, augmented reality, and games offers interesting educational possibilities for outdoor historical sites. One particular challenge is that no two locations are alike. Some historical sites—many battlefields for instance—are not much more than a field, while other places, such as Harpers Ferry, consist of many intact structures, both original and reconstructed. How can students navigate these sites and connect with the "ghosts" that reside there? How can educators make the invisible visible (Robinett, 1991)? Augmented reality provides the means of enhancing the environment with added sensory elements that can bring some of those ghosts to life, and it was the possibility of more fully exploring John Brown's role in Harpers Ferry's rich history through that medium that I started considering the idea of designing an AR mobile game. As an educator, I have often heard students complain about being tethered to the group while on field trips, unable to move at will. Ranger Dennis Frye has observed the same frustration, noting how much students dislike "being led around on a leash" (D. Frye, personal communication, April 2013). The medium of an augmented reality game would certainly afford students the mobility they craved, and I was curious to find out what the pedagogical value might be.

Starting by questioning how students make meaning out of such an experience, I began the design process. The game was titled *Insurrection*, a reference to Brown's attempted takeover of the federal armory and arsenal. The game's development was initially hindered by the lack of some necessary frameworks to provide a clearer direction for the game's design, and these needed to be established before proceeding. The first, a conceptual framework, outlines the role of immersion and engagement as they relate to "space" and "place" for outdoor historical sites. The second, an augmented reality design framework for historical sites, details the kind and degree of appropriate augmentation. With those frameworks in place, the game took shape over a 6-month period using the ARIS platform (ARIS, 2010). A qualitative research study to explore meaning-making was conducted consisting of surveys, interviews, and field observations, and the game was play tested in April 2013 with 18 eighth-grade students from Harpers Ferry Middle School. Although limitations of the study included a short, 1-day testing period and a small sample size, analysis of the results suggested that players did find the enhanced experience an edifying one, based on not only their enjoyment of the game but also on their ability in post interviews to demonstrate a greater understanding of the complexities of the raid itself. I hope that other educators who are considering augmented reality gaming in informal learning environments such as national parks will find these results encouraging.

FRAMEWORKS INFORMING GAME DESIGN AND GAME PLAY

History Education

The first big challenge when designing a historical game is knowing where to begin. As Karen Schrier (2014) points out, "one of the driving questions in history education is what types of content, skills, and practices it should include" (p. 75). The focus could be the simple learning of facts, names, dates, as well as a set narrative, although the belief that history is nothing more than this can lead to an "intellectual passivity" (McCall, 2011, p. 9), where students will not bother to reason through the available evidence and construct their own interpretations of the past. However, another approach to history education emphasizes historical thinking as an exercise where one constructs meaningful interpretations of the past by examining evidence and interpreting perspectives while acknowledging that many other interpretations are possible (McCall, 2011; Schrier, 2014). Further, as Schrier (2014) notes, history educators may choose to employ a narrower focus by concentrating on the people and their struggles, emphasizing resources or technologies, or guiding students in trying on other cultural and social contexts to help them relate to the issues and decisions of the past, something Schrier refers to as historical empathy. When students can reconstruct "other people's beliefs, values, goals, and attendant feelings" (Ashby & Lee, 1987, p. 63), they can better understand the past by putting themselves "in the mindset of someone in history" (Schrier, 2014, p. 82).

Broader Learning Theories

Outdoor historical spaces present their own unique set of challenges since they are neither classroom settings with textbooks nor strictly curated environments such as museums. What is the value in visiting? A constructivist approach as well as the concept of situated cognition guided how I theorized my research.

Constructivism and Communities of Practice

Jean Piaget's constructivist theory centers on how learners build ideas and meaning. Students construct their own meanings and create their own interpretations by interacting with their environment and "mastering authentic tasks in meaningful, realistic situations" (Dede, 2008, p. 51). Even if there are established, "agreed-upon theoretical constructs" (Wiggins & McTighe, 2006, p. 92) in a field of study, "didactic teaching of *the* interpretation will mislead students about the truly arguable nature of all interpretation" (Wiggins & McTighe, 2006, p. 92). Another related learning framework applicable here is the idea of communities of practice (Lave & Wenger, 1991) where students work collaboratively, engaging in a "process of collective learning in a shared domain" (Wenger-Trayner & Wenger-Trayner, 2015, p. 1) to solve authentic problems.

Situated Cognition

Situated cognition is the idea that "information exists not prior to, but emerges from and is a function of, the organism-environment relation" (Roth & Jornet, 2013, p. 464). In other words, cognition is connected to the interaction between agent and physical environment; and how environment, situation, and context present themselves to the agent, and how cultural resources are utilized, is a vital part of the learning process (Bruner, 2009; Roth & Jornet, 2013; Schrier, 2006). Psychologist Jerome

Bruner, in the context of meaning making, views the mind as actively situating experiences within their cultural contexts.

> Meaning making involves situating encounters with the world in their appropriate cultural contexts in order to know "what they are about." Although meanings are "in the mind," they have their origins and their significance in the culture in which they are created. (Bruner, 1996, p. 3)

Bruner (1990) further maintains that meaning, and the processes and transactions involved in the construction of meaning, is the central concept of human psychology. This construction is a reciprocal one where the world in which our acts are situated influences our desires and beliefs. Those desires and beliefs in turn influence the meanings we create. Since desires and beliefs are different for each individual, they may lead some to find meaning in contexts that others might not (Bruner, 1990). In essence, no observation is independent of the effects of the observer's experiences, beliefs, personal values, and expectations. Nor, as Bruner (1996) asserts, is meaning making free of feelings and emotions, for "surely emotions and feelings are represented in the processes of meaning making and in our constructions of reality" (p. 12).

AR Games for Historical Sites

The advantages of taking students away from history books and placing them in situ are their experiencing and interacting with an actual historical space and playing a game that is highly specific to a particular location. Such a game can incorporate a great deal of detail about the real-world location, allowing tight connections between player, game, and world (Klopfer, 2008). The biggest obstacle is the potential lack of familiarity with the site prior to the visit. Knowing an area well beforehand allows the successful applying of preexisting knowledge and feelings about a locale to a complex problem. The resulting understanding can then be transferred back to the real world (Klopfer, 2008). Where unfamiliarity with a site precludes an emotional connection, augmented reality can help. If students are involved in a game and do not care enough about the physical space through its "real-world meaning" (Klopfer, 2008), it is likely that the connection can be enhanced through the virtual component of AR (Klopfer & Squire, 2007; Rogers et al., 2004). The affordances of AR—its blend of actual and virtual elements, and its situated nature that allows students to learn in context—coupled with the affordances of good games, such as the freedom to make choices and direct the narrative of the game play (Klopfer, Osterweil, & Salen, 2009), suggest a strong likelihood that AR can be an extremely successful gaming medium, and that AR gaming itself—with the soft skills that gamers employ such as problem-solving, strategic thinking, systems thinking, and collaboration—will be an effective 21st century learning strategy (Federation of American Scientists, 2006; Gee, 2007; Klopfer, 2008).

Immersion and Engagement

Inherent in any discussion about technologically mediated experiences, including games, are the concepts of immersion and engagement. Immersion grew out of virtual reality research and refers to "the extent to which a person's cognitive and perceptual systems are tricked into believing they are somewhere other than their physical location" (Patrick et al., as cited in Brown & Cairns, 2004). In the case of visits to historical sites, however, the physical location is exactly where educators want learners to be. It is not the "somewhere," but the time—the "somewhen"—educators should like to impose on students' perceptions. Engagement occurs when students interact with that "when" on a cognitive and emotional level, when, as Murphy (2011) asserts, "you are actively involved in trying to work through a problem or puzzle" (p. 6). Further, Price, Jewitt, and Sakr (2016) indicate that "in the context of history learning, digitally augmented experiences of place can 'bring closer' aspects from the past, enabling student engagement in location with different time periods simultaneously" (p. 345).

CONCEPTUAL FRAMEWORK

Price et al. (2016) define "place" as a sense that is "developed through embodied experiences of a space that is imbued with cultural and social associations." Figure 8.1 illustrates a conceptual framework for *Insurrection* and for outdoor historical sites more broadly. Central to this framework is the concept of "space," or in Price et al.'s (2016) view, the physical location without the cultural and social associations. A number of elements, as illustrated in this framework, contribute to students' perceptions of that space. The first is the fact that this space is an informal learning environment away from their school—a field trip—and one that students may or may not be familiar with. Even if they are familiar with the environment itself, students still may not have any knowledge of the people or events that make that space an important one. However, what the field trip offers is an opportunity for the mind and body to work

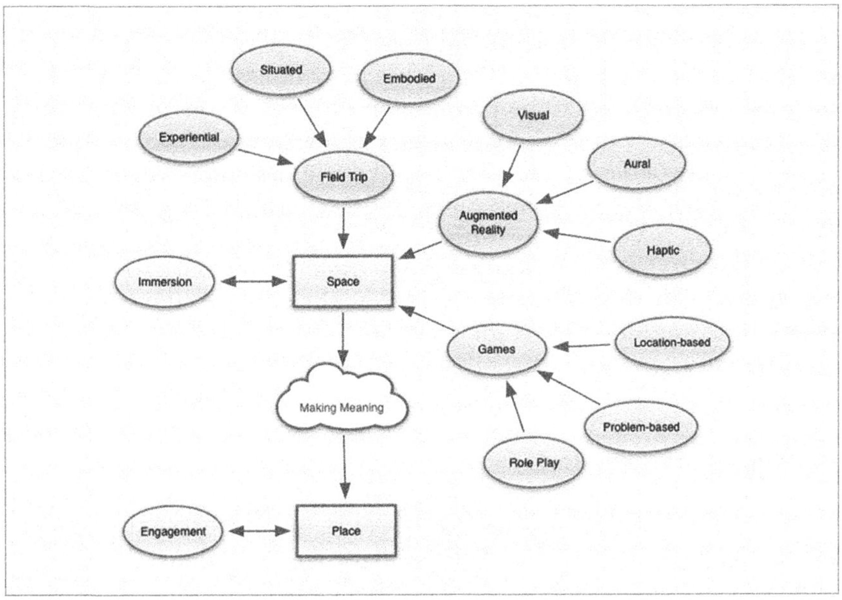

Figure 8.1. Conceptual framework.

together as it interacts with the physical environment. As Roth and Jornet (2013) emphasize, "cognition is situated and embodied" (p. 464).

The second element contributing to the space is any added digital augmentation. This could include any enhancements in any of the three main sensory modalities: visual, aural, and haptic. Depending on the size and physical layout of the space, and the degree of material culture that exists, augmentation may be light or heavy and consist of only one type of sensory augmentation or some combination of all three. The third aspect of this framework is any activity that students might be involved in while visiting the site. For the purpose of this study, the activity is a mobile game. Although students could approach the space in a number of ways such as guided tours or solo exploration, a game offers a location-based and problem-based experience by incorporating an actual location, and through digital augmentation, the characters, events, and stories that make that site significant. Students have the opportunity to inhabit specific roles, solve puzzles, and work through challenges, perhaps even change the outcome of historical events, something McCall (2016) refers to as "counterfactual narratives" (p. 525).

These three aspects, the on-site visit, any added digital augmentation, and the gaming format, contribute to the immersion in a space, a physical location devoid of any social or cultural associations. From an educator's perspective, however, students will hopefully experience more than just physically entering a space. History teachers and National Park Service rangers work very hard to prepare students to visit sites by showing films, assigning chapters in textbooks, or giving interpretive talks in order to convey a sense of the people, events, and stories that make these sites historically important. Certainly, one of the education objectives of the National Park Service is to "further the view of national parks as classrooms" (Pitcaithley, 2002). Rangers hope students will make some kind of connection with their sites that will allow them to see those spaces as places with meaning, and with social and cultural contexts that make them unique.

Designing *Insurrection*

Designing *Insurrection* for Harpers Ferry National Historical Park was a multifaceted project. There were three basic considerations: (1) the kind and degree of augmentation, (2) the specific platform, and (3) the game play itself. The augmentation of an outdoor historical site requires a close look at the qualities of the site itself. No two sites are alike; therefore, no one style of augmentation is appropriate. A site might include primarily natural features like many battlefields; or it might include copious material remains such as one might find in the ruined abbeys and monasteries scattered throughout England. A site might include an entire town, perhaps, with authentic or reconstructed elements, or it might be some combination of all of these.

Harpers Ferry's specific qualities feature a unique blend of natural and cultural resources. Being situated at the confluence of the Potomac and Shenandoah rivers, Harpers Ferry was a center of industry due to the easy access to water power. Mills of various kinds existed there at one time, and most importantly, a federal armory, arsenal, and rifle factory took advantage of the power the rivers had to offer. The strong military presence, the Baltimore and Ohio (B&O) railroad, and the Chesapeake and Ohio (C&O) Canal also contributed to Harpers Ferry as a bustling industrial center. Because of its strategic importance in a mountainous region on the borders of three states, the town was highly contested during the Civil War, changing hands between Union and Confederate forces eight times, and was the site of the largest surrender of Union troops. John Brown took an interest in Harpers Ferry as a site for a raid, both for its easy access to weapons for his desired slave uprising and for the mountainous terrain he hoped would provide him the natural fortress he needed to succeed (Frye, 2009; NPS, 2009; PBS, 2000).

John Brown abhorred slavery, and though his goals were always lofty, he sometimes resorted to brutal methods to further his antislavery cause. After many months of planning and preparation, Brown and his band of 18 men crossed the Potomac from Maryland into then-Virginia (now West Virginia) on October 16, 1859. They surprised the watchman and easily captured him, and in short order took over the armory, the arsenal, and the somewhat distant rifle factory. The basic plan was to confiscate weapons, arm the Southern slaves, and lead them north to freedom through the mountains. The plan unraveled once neighboring militias and the US Marines arrived, and by October 18, it was all over very quickly. Barricaded in the armory fire engine house, John Brown made his last stand as the Marines broke through the door and killed or captured the remaining insurgents. Brown himself was captured and later executed that December (Frye, 2009; NPS, 2009; PBS, 2000; Varon, 2008); and though the raid seemed to be a failure, Brown became a martyr for the North and a terrorist in the South, splitting the country apart (Frye, 2009; Varon, 2008). The question of the success or failure of the raid, the motivations of the men who joined John Brown, and the moral ambiguity of Brown himself provided ample story fodder for *Insurrection*.

Design Framework

In 1994, Paul Milgram and Fumio Kishino developed a Virtuality Continuum that ranged from the real environment on one end to a complete virtual environment on the other. In between lay AR toward the real end of the continuum, and augmented virtuality toward the virtual end. Expanding on this framework, MIT researcher Dr. Eric Klopfer modified the augmented reality subset to further develop the relationship between the real and virtual elements (Figure 8.2). Klopfer's illustration shows that as the degree of real-world input decreases, the degree of virtual input increases, suggesting that environments with a rich material culture would not benefit from a heavy degree of augmentation since such augmentation would detract from the real-world input; however, environments with a dearth of such resources—many battlefields, for instance—might benefit greatly from heavier augmentation. In order to be more specific to historical sites, this framework can be modified to include the three main sensory modalities—visual, aural, and haptic—that are capable of augmentation, as well as the inherent challenges and variable characteristics of historical sites. Figure 8.3 illustrates the elements of an enhanced experience of a historical site—physical, historical, and virtual—along with the heart of the framework, the player's imagination, which pulls all the elements together and offers its own contribution to the experience.

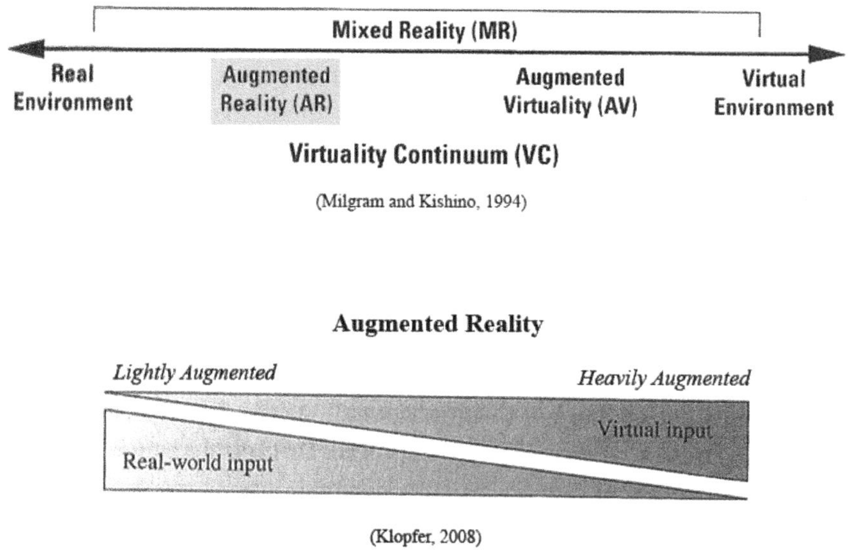

Figure 8.2. Augmented reality on the virtuality continuum.

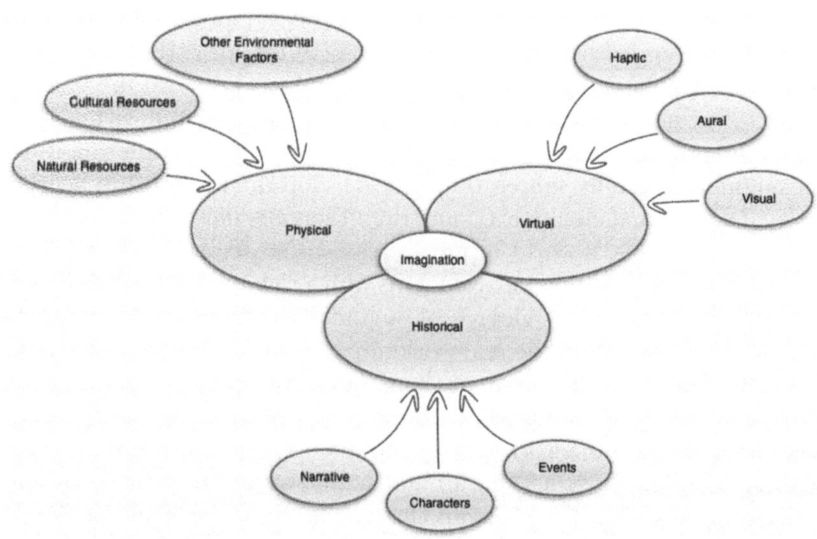

Source: Adapted from Stapleton, Hughes, Moshell, Micikevicius, and Altman (2002).

Figure 8.3. Augmented reality design framework for historical sites.

The physical portion of this framework is generally fixed. Each site will have its own unique terrain and natural features, unique cultural resources, and the possibility of additional environmental factors as well. Although these natural and cultural resources could change, such change would be likely to happen—barring catastrophe—only over a very long period of time. Other environmental factors such as traffic or excessive noise are a bit more unpredictable but are generally consistent. In the case of Harpers Ferry National Historical Park, it is a town in its own right with residences and businesses sprinkled among the museums and 19th century buildings. Its natural resources include mountainous terrain; the Potomac and Shenandoah rivers, which converge at this point; and the Appalachian Trail, which runs directly through the town and brings through large numbers of hikers. Other environmental factors include several roads that run through the lower town, making traffic a very real safety concern. Because there is such a rich array of cultural and natural resources at Harpers Ferry, as well as some environmental factors that warrant caution, any augmentation should be light and work to enhance rather than distract.

The next facet in the framework is the historical. This aspect is also generally fixed. The stories of a place, including the actual events and the people who lived through them, fall into this category. They can range from a single event to a multitude of stories (keeping in mind that some of these stories may never be fully known or may change over time as new information comes to light). Harpers Ferry is a space with multiple storylines, though *Insurrection* focuses on only one: that of John Brown's raid.

The third aspect of designing for historical sites is the virtual input, and it is this aspect that is dependent on the other two. Klopfer's (2008) framework recommends a lightly augmented experience based on the degree of real-world input, but it does not address what sensory modalities to emphasize. With traffic being such a concern at Harpers Ferry, a heavily visual experience would likely be too distracting and unsafe. Instead, a digitally enhanced experience that relies more on the aural and/or haptic would be a better design focus, with any visual augmentation being secondary. However ideal this might be, cost-effectiveness and ubiquity favor smartphones for handling the augmentation.

Building the Game

Ironically, the biggest obstacle to moving forward with *Insurrection* was simply figuring out what the game goals were going to be. For a long time, recreating the raid and having players learn about it by playing raiders and "living through it" seemed like the logical way forward. However, after weeks of trying, there was absolutely no progress. It became clear in a very palpable way that the players' lack of choice completely inhibited the design process. A second iteration of the game put the players in the roles of townspeople who would observe the raid unfolding. Perhaps they would encounter Brown and the raiders. Perhaps they might choose to help or hinder the plot. That narrative went a little further initially, but then it hit a brick wall again. There was a little bit of choice, but not nearly enough, and the role of townspeople put the players in the undesirable position of being on the outside looking in.

McCall (2016) discusses historical games at length and, especially, the role of choice. Historical games encourage players to consider the context as well as the affordances and constraints of a historical problem space. Having the freedom to make choices, especially choices similar to those encountered by the historical characters, can be a valuable experience. Squire (2011) also indicates that one of the core features of local games that drives learning is emotionally compelling challenges, and overcoming challenges requires agency. Even if players make different choices from their historical counterparts, the experience of the choice and its consequences can be, as McCall (2016) asserts, a valid way to teach history.

> Essentially, historical games generate the material of counterfactual narratives. The outcome of a simulated situation, if the simulation has enough complexity and offers the player sufficient choice, will often not match what historical evidence suggests actually happened. Therefore the player can consider the differences between the simulated events and the historical evidence. (McCall, 2016, p. 525)

Stepping into this idea of counterfactual narrative, the third iteration of the game had a simple goal: sneak into town, find six weapons caches, and get out without being caught. This is not too far removed from what John Brown broadly intended to do, but instead of making the exact choices he and his men did, players encountered similar decision points and were free to make their own choices. Instead of following Brown's precise path, the players decided which direction to go and how to interact with the characters they encountered.

Of course, there were many aspects of the game that mirrored the historical raid. The players were themselves raiders and part of John Brown's band. In order to get a sense of the physical space Brown had to navigate, and to understand why Frederick Douglass felt that Brown and his men were walking into a "perfect steel-trap" (NPS, 2009, p. 28), players needed to enter the town in the same way Brown did: by starting on the Maryland side of the Potomac River and walking across the narrow railroad bridge. That original bridge (today a ruin) was the way into and the way out of Harpers Ferry, and it would not take much to cut off that route of escape. Walking across a more recent bridge, players encountered many of the same people that Brown and his men did. On the railroad bridge, for instance, was a virtual William Williams, the watchman, on the lookout just as he was in 1859. Other characters appeared as players came within proximity of them, and historical decision points emerged as players interacted with these characters. Certain locations that were significant to the raid, such as the armory and arsenal, were required stops. However, the path through the game and the decisions the players made were entirely up to them.

Designed to take approximately 2 hours to play, *Insurrection* was built in three levels and contained a total of 17 virtual characters. It featured six additional locations with no characters present, and 13 virtual but authentic artifacts. Wherever possible, characters and artifacts were integrated into the game using actual images from the period (see Figure 8.4). Each level contained two of the six weapons caches the players needed to find.

Players learn at the outset of the game that two of the weapons caches can be found at the armory and at the arsenal. Coming across the bridge into Harpers Ferry's lower town, players encounter a virtual watchman and have their first choice to make. Depending on what they choose, the watchman will either let them go peacefully or regard them warily, which will garner them a suspicion point. Once past the watchman, players can choose their path through level one. Many of the obstacles players encounter, and the choices they face, are based on actual events and are thus very important to the game's design. They encounter an armory gate that they cannot open without a crowbar. Where can they get one? They

Figure 8.4. Screenshots from *Insurrection*.

discover that a train coming into the old station could contain passengers who might alert the authorities. Do they stall the train or not? If they do stall the train, the baggage man runs off to tell someone. Do players shoot him or let him go? If suspicion points run too high, an alert lets players know the townspeople are watching them. If they visit the tavern, they will discover that buying a round of drinks for the townspeople eliminates one suspicion point. However, they only have a limited amount of money and can only resort to this on a few occasions.

By visiting all the game locations and using the virtual resources available to them, and by paying attention to what the virtual characters reveal during the course of the game, players ideally get a sense of where park sites are located (or used to be located in 1859). They discover that the names they have heard about in history class actually have faces and opinions. They realize that John Brown's raid was not just about one man and his ill-advised scheme, but that the raid was a long-planned operation involving many people. At the end of it all, through class discussion, they can ideally demonstrate that this event involved far greater complexity than they realized. In the next section, I describe the study in which a group of middle school students agreed to visit Harpers Ferry National Historical Park and play test *Insurrection*.

Methodology

Participants in this study were 18 eighth graders in a single middle school history class at the nearby Harpers Ferry Middle School. The school, and this particular teacher, had a very close relationship with Harpers Ferry National Historical Park, and the students went there very often for various activities. The students were Caucasian, evenly split between boys and girls, and exhibited a wide range of academic ability and technological expertise. Data types consisted of two teacher questionnaires, administered prior to and immediately following the day of play testing. In addition, there were three sets of student interviews. The first was to establish their prior experiences with field trips as well as their basic knowledge of John Brown. The second took place the day before play testing to establish their expectations, and the third recorded their reflections once the field trip was over.

The first round of interviews established the familiarity the students had with field trips in general and Harpers Ferry in particular. All of them had been there at least once, most of them many times, either with family or just to hang out with friends. Interestingly, most of them knew surprisingly little about John Brown's raid, despite the fact that the class had just completed a chapter on the subject and had watched a short film. Rather than a meticulously planned operation, their vision of Brown and the raid was more of an impulsive, lone-wolf scenario lacking any complexity whatsoever.

During the second round of interviews, the students expressed equal parts excitement and trepidation about what might happen the following day. Several of them were self-described technophobes and were very nervous about having to handle smartphones. They were relieved to hear that students would be working in pairs, and no one would be left to flounder on his or her own. Other students were on the opposite end of the spectrum and were very relaxed and eager for the next day. Though the promise of a unique educational experience might have been the reason, it is also possible the students were looking forward to a morning of missed classes.

Prior to the field trip, the teacher divided the 18 students into nine teams of two. Once at the park, each team received an iPhone 4s and a USB audio recorder on a lanyard. One team member carried the phone while the other had the audio recorder around his or her neck. Students had no exposure to the software prior to the day of testing. Five dedicated observers were present to follow particular teams around for the duration of the game and to record field observations. These might include any technical difficulties the students were having, any excitement or frustration they might express, or any insights they might have about the game play. These observers were briefed before the students arrived to ensure they understood that they were only to record observations; they were not to offer any help to the players. Other ad hoc observers, most notably from the National Park Service, wandered from team to team, or stayed in the center of town and observed as teams passed back and forth. All notes were collected afterwards, and the National Park Service rangers offered feedback during a debriefing once the play testing was over.

Despite issues such as technical glitches, inexperience with the game software, and typical device constraints such as screen glare and limited battery life, posttesting interviews contained ten references to the field trip as "good," six references to the field trip as "more fun than other field trips," and three references to the field trip as "one of the best field trips." The novelty effect could account for some of this, of course, but what was undeniable was that those who committed themselves to playing the game (and there was one team that did not) responded outside the lone-wolf scenario with these and similar comments when asked what they learned about John Brown and his raid as a result of playing *Insurrection*:

> I never knew he was a religious man.
>
> I didn't know he planned this with so many people.

> I never realized how many people it actually took ... there was a lot of stuff beforehand to set it up.

> You could see all the ways it could go wrong. Think about it from John Brown's point of view ... how easy it would be for the soldiers to spot you, shoot you, then the whole plan is messed up.

While it would be foolish to jump to any conclusions based on one morning's gaming with a middle school history class, the post interviews certainly suggest that some cognitive benefit was derived from experiencing the John Brown "counterfactual narrative" (McCall, 2016, p. 525) in situ with augmentation that allowed the students to interact with virtual characters and artifacts from the past and, through playing, graft that social and cultural context onto the space to make it a "place." Based on axial and selective coding of the interviews and the field data, agency and choice were clearly important to the students. As a result of having the freedom to move around the park, teams were able to make their own decisions about where to go, about how to interact with characters, and to make choices about puzzles and decision points. What was most exciting to them can be summed up in one student's comment: "I loved how what I chose affected the game." Players found the role playing to be fun, and the game feedback was validating when, for example, a team found a weapons cache or leveled up, and players discovered new places and learned new things about game locations they already knew existed. As an added bonus, playing *Insurrection* opened up a whole new world of ideas related to the game design. Almost every team had suggestions for improvements, usually involving an increase in the danger and in the number of obstacles.

CONCLUDING THOUGHTS

Our nation's national historical parks are rich cultural resources that should be accessible to all who are interested in visiting. The more tools educators have to access that history, the better we can serve all students, whether they are formal classroom students or the casual armchair historian. Augmented reality games for historical sites are not meant to take the place of guided tours, ranger-led discussions, visitor center displays, or on-site museum exhibits. Rather, the augmented reality game is meant to be another tool in the toolbox for educators seeking to make these sites more approachable for everyone, especially in the event that budget cuts leave parks too short-staffed to adequately provide the educational services they would like to. If educators keep the site and the stories of a particular site in mind, allowing the site to dictate the augmentation and the stories to direct the gameplay—without getting too caught up in the need to relive history—they can create some truly rewarding experiences.

In my own exploration of Harpers Ferry, the medium of an augmented reality game gave students the opportunity to navigate the past with at least one hand on the tiller. We as educators have the fascinating opportunity to allow these counterfactual narratives to come to life, and in so doing, choose which aspects of the historical narrative to keep fixed and which to let students reimagine. *Insurrection* leaves intact the importance of the topography of Harpers Ferry, the primary characters, various 19th century artifacts, and important decision points. Within those parameters, students are free to create their own narrative as they inhabit this historical space and solve problems against the backdrop of the smoldering powder keg of Virginia in 1859.

REFERENCES

ARIS. (2010). Augmented Reality and Interactive Storytelling [Computer software]. Retrieved April 15, 2017, from arisgames.org

Ashby, R., & Lee, P. (1987). Children's concepts of empathy and understanding in history. In C. Portal (Ed.), *The history curriculum for teachers* (pp. 62–88). London, England: Falmer.

Brown, E., & Cairns, P. (2004, April). A grounded investigation of game immersion. *CHI '04 Extended Abstracts on Human Factors in Computing Systems*, Vienna, Austria.

Bruner, J. (1990). *Acts of meaning*. Cambridge, MA: Harvard University Press.

Bruner, J. (1996). *The culture of education*. Cambridge, MA: Harvard University Press.

Bruner, J. (2009). Culture, mind, and education. In K. Illeris (Ed.), *Contemporary theories of learning* (pp. 159–168). New York, NY: Routledge.

Dede, C. (2008). Theoretical perspectives influencing the use of information technology in teaching and learning. In J. Voogt & G. Knezek (Eds.), *International handbook of information technology in primary and secondary education*, (pp. 43–62). New York, NY: Springer.

Federation of American Scientists. (2006). *R&D challenges in games for learning*. Washington, DC: The Learning Federation.

Frye, D. (2009). Purged away with blood: John Brown's war. *Hallowed Ground, 10*(3), 16–20.

Gee, J. (2007). *What video games have to teach us about learning and literacy*. New York, NY: Pallgrave/McMillan.

Klopfer, E. (2008). *Augmented learning: Research and design of mobile educational games*. Cambridge: MIT Press.

Klopfer, E., & Squire, K. (2007). Augmented reality simulations on handheld computers. *The Journal of the Learning Sciences, 16*(3), 371–413.

Klopfer, E., Osterweil, S., & Salen, K. (2009). *Moving learning games forward: Obstacles, opportunities, and openness* [White paper]. Retrieved November 15, 2016, from MIT Education Arcade website: http://education.mit.edu/wp-content/uploads/2015/01/MovingLearningGamesForward_EdArcade.pdf

Klopfer, E., Osterweil, S., & Salen, K. (2009). Moving learning games forward: Obstacles, opportunities, and openness. *The education arcade*. Cambridge, MA: MIT Press.

Lave, J., & Wenger, E. (1991). *Situated learning: Legitimate peripheral participation.* Cambridge, England: Cambridge University Press.

McCall, J. (2011). *Gaming the past*. New York, NY: Routledge.

McCall, J. (2016). Teaching history with digital historical games: An introduction to the field and best practices. *Simulation & Gaming*, *47*(4), 517–542.

Milgram, P., & Kishino, F. (1994). A taxonomy of mixed reality visual displays. *IEICE Transactions on Information and Systems*, 77(12), 1321–1329.

Murphy, C. (2011). Why games work and the science of learning. *Alion Science and Technology*. Retrieved from http://www.goodgamesbydesign.com/Files/WhyGamesWork_TheScienceOfLearning_CMurphy_2011.pdf

National Park Service. (2009). *John Brown's raid handbook*. Harpers Ferry, WV: National Park Service History Series, Harpers Ferry Historical Association.

Public Broadcasting Service. (2000). John Brown's holy war. *American Experience*. United States: PBS Home Video.

Pitcaithley, D. T. (2002). National parks and education: The first 20 years. *History E-Library. National Park Service*. Retrieved from https://www.nps.gov/parkhistory/resedu/education.htm

Price, S., Jewitt, C., & Sakr, M. (2016). Embodied experiences of place: A study of history learning with mobile technologies. *Journal of Computer Assisted Learning*, *32*, 345–359.

Robinett, W. (1991, Fall). Electronic expansion of human perception. *Whole Earth Review*, 16–21.

Rogers, Y., Price, S., Fitzpatrick, G., Fleck, R., Harris, E., Smith, H., ... Weal, M. (2004, June). Ambient wood: Designing new forms of digital augmentation for learning outdoors. In *Proceedings of the 2004 conference on Interaction design and children: Building a community* (pp. 3–10). New York, NY: Association for Computing Machinery.

Roth, W., & Jornet, A. (2013). Situated cognition. *WIREs Cognitive Science, 4*, 463–478. doi:10.1002/wcs.1242

Schrier, K. (2006). *Using augmented reality games to teach 21st century skills.* Paper presented at the ACM SIGGRAPH 2006 Conference, Boston.

Schrier, K. (Ed.). (2014). Using digital games to teach history and historical thinking. In *Learning, education and games: Vol. 1. Curricular and design considerations* (pp. 73–91). Pittsburgh, PA: ETC Press.

Squire, K. (2011). *Video games and learning: Teaching and participatory culture in the digital age.* New York, NY: Teachers College Press.

Stapleton, C., Hughes, C., Moshell, M., Micikevicius, P., & Altman, M. (2002). Applying mixed reality to entertainment. *Computer, 35*(12), 122–124.

Varon, E. R. (2008). *Disunion! The coming of the American Civil War, 1789–1859.* Chapel Hill, NC: The University of North Carolina Press.

Wenger-Trayner, E., & Wenger-Trayner, B. (2015). Communities of practice: A brief introduction. Retrieved from http://wenger-trayner.com/wp-content/uploads/2015/04/07-Brief-introduction-to-communities-of-practice.pdf

Wiggins, G., & McTighe, J. (2005). *Understanding by design*. Alexandria, VA: Association for Supervision and Curriculum Development.

PART III

MOBILE LEARNING IN PRE-K–12 EDUCATIONAL SETTINGS

CHAPTER 9

PIECING THE PUZZLE TOGETHER

Building a Dialogue for Engaging Multiple Disciplines in Inquiry-Based Mobile Learning Through Professional Development

Cynthia C. Minchew Deaton and Sandra M. Linder
Clemson University

Benjamin E. Deaton
Anderson University

This chapter opens a dialogue for conceptualizing mobile learning in K–12 education through the lens of inquiry-based pedagogical practices. As mobile learning has become common in educational settings (Kukulska-Hulme, 2010), professional development providers may struggle to support teachers as they move toward integrating mobiles to support learner-centered practices. How can we, for example, conduct individualized professional development to meet teachers' contextual needs when focusing on broad concepts such as integrating mobiles? How can we address teacher contexts when mobile learning professional development experiences could contain participants from varying content disciplines,

Mobile Learning: Perspectives on Practice and Policy
pp. 191–209

varying grade levels, and with varying pedagogical expertise? And most importantly, how can we move teachers away from teacher directed and managerial use of mobiles in the classroom and toward learner-centered mobile integration? We propose the use of inquiry as a framework for conducting professional development and guiding pedagogical principles in implementing mobile learning regardless of grade level or content area. In this chapter, we will discuss a 2-year professional development experience that integrated mobile learning with inquiry-based pedagogical practices to promote critical thinking and learner-centered approaches across disciplines. As we discuss this experience, we will present findings that illustrate how this professional development model encourages learner-centered instruction, innovative uses of technology, and increased teacher self-efficacy for integrating mobile learning.

INTEGRATED APPROACH TO MOBILE LEARNING PROFESSIONAL DEVELOPMENT

A primary goal of professional development is the need for teachers to engage in continued learning that is related to the contextual issues they face as teachers in their classroom everyday (Hoyle, 1982). In today's classrooms, learning context is heavily laden with the notion of using mobiles for teaching and learning. As new national standards have been adopted over the last 5 years (e.g., technology standards from the International Society for Technology in Education, Next Generation Science Standards), so have initiatives focused on finding more engaging and innovative ways to encourage learning. The integration of mobiles is promoted as an avenue for engaging students in content through student-centered learning. The impact of mobiles on established classroom practices, however, should be considered prior to introduction (Misfud, 2014) in that their introduction influences the normal network of learning. In order to encourage teachers to use them effectively with their students, teachers need more than a brief introduction on how mobiles can replace or supplement existing technology. That said, effective models for supporting teachers to actively and effectively use mobiles for teacher and student driven activities is vital.

Professional development experiences should be relevant to teachers' practices, focus on their daily activities, and address student and teacher concerns (Guskey, 1995; Loucks-Horsley, Stiles, Mundry, Love, & Hewson, 2010), and should marry mobile learning initiatives with effective practices for implementing new national standards. To effectively support teachers in understanding and integrating mobiles, professional development needs to provide opportunities for collaboration, modeling, practice, and reflection while embedding the learning in the local context and providing extensive support (Loucks-Horsley et al., 2010; Posnanski, 2002).

Considering mobile use is driven by pedagogical considerations (Ozdamil, 2012), an inquiry-based focus can serve as a thread to support multiple grade levels and disciplines in the same professional development experience. Inquiry-focused professional development can provide teachers with agency, which will increase teacher learning (Loucks-Horsley et al, 2010) as teachers determine how to best integrate inquiry-based instruction and mobiles for their discipline. The driving process of inquiry instruction and inquiry-focused professional development is the notion of problem solving. Instead of receiving didactic instruction (in either a classroom setting or a professional development session) that requires passive engagement from participants and a strong focus on the role of the teacher or professional development facilitator as an expert, inquiry-based approaches require the teacher or facilitator to act as a guide while students or professional development participants collaborate to work toward a goal. Essential features of inquiry include increased collaboration and goal-focused discourse; increased opportunities for critical thinking, reasoning, and developing strategic competence; and increased opportunity for reflection and making real-world and discipline-based connections (National Research Council, 2000). Mobiles can enhance learner-centered inquiry instruction in a variety of ways. From a teacher point of view, mobiles can support assessment of student understanding and can support recording and analyzing this data to make instructional decisions. From a student point of view, mobiles can act as tools for engaging in a problem-solving task, as a mechanism for communicating understanding, and as a resource for increasing strategic thinking by generalizing concepts through a variety of applications. Through teacher and student generated products resulting from inquiry-based lessons, mobiles become tools to support communication, creativity, collaboration and other 21st century skills.

DESIGNING AND IMPLEMENTATION OF AN INTEGRATED PROFESSIONAL DEVELOPMENT

Our professional development model is based on the situated learning epistemological stance. Situated learning stresses that what is learned is connected to the context in which it is learned (Anderson, Reder, & Simon, 1996; Brown, Collins, & Duguid, 1989). To use and understand tools of practice, the learning of the tools must be linked to the context in which they are used (Brown et al., 1989; Hedegaard, 1998). Mobile learning theories align with situated learning in that they emphasize context, connectivity, authenticity, and personalization (Crompton, 2013; Kearney, Schuck, Burden, & Aubusson, 2012). The notion of personalization and authenticity employs motivational and sociocultural theories through customization and agency given to learners (Kerney et al., 2012).

We provided professional development through a university-based Center of Excellence focused on mobile learning (CML), which supports and promotes innovation in teaching and learning through the design and development of learning experiences that integrate technology. The CML's main focus was to provide K–12 teachers with support in integrating mobiles into their classroom. To begin supporting K–12 teachers in mobile learning, we partnered with a school district classified as high-needs in regards to student socioeconomic status and student performance data identified by the state's annual yearly report. Our district partner was also adopting a new mobile learning initiative. Namely, the district was implementing a 1:1 adoption of MacBook's for Grades 6–12 and a 1:1 adoption of iPads for Grades 2–5. Grades pre-K–1 had access to classroom sets of iPads within their school. Additionally, one member of the CML's leadership team (which included the chapter's third author) had extensive experience in both using mobiles and providing professional development to others in using mobiles in the classroom. Using the districtwide mobile learning initiatives and annual yearly report, we developed a multiyear plan for scaffolding mobile learning support that supplemented support provided by the district.

A 3-year mobile learning professional development was conceptualized to engage teachers of all grade levels and subject areas. We grounded the focus of this 3-year model in discussions with the partnering school district related to student and teacher needs in addition to pedagogical considerations for engaging multiple disciplines and multiple grade levels in a single professional development model. The primary goal was to provide practicing teachers with tools and strategies to promote critical thinking, collaboration, communication, and creativity in their classrooms through the use of mobile learning. In doing that, we aimed to prepare teachers at our partner school district to effectively integrate mobiles for teaching and learning. We used a three-stage approach to implement a professional development model that expanded the number and grade level of participants from year to year as the topics and emphasized technologies changed. The first year of CML profession development (CML PD) focused solely on working with middle grade teachers. CML PD activities supported participants in developing an understanding of inquiry-based instruction, using standard Apple software (e.g., Keynote [Apple Inc., 2017c]), Photos (Apple Inc., 2017d) found on their mobiles, and learning Comic Life (Plasq, 2017) and Inspiration (Inspiration Software,

Inc., 2017). In the second year, the CML PD was opened to secondary teachers. Activities for Year 2 emphasized the use of inquiry-based teaching and cross-disciplinary skills (e.g., critical thinking, creativity, collaboration and teamwork, communication, and metacognition), which were spotlighted in state initiatives. In Year 2, CML PD facilitators further emphasized the common set of tools that were encouraged in Year 1. Participants were also asked to begin integrating outside content, such as materials from Khan Academy (Khan Academy, 2017) and Illuminations (National Council of Teachers of Mathematics, 2017). Those web-based resources provided teachers with materials to support their teaching and learning goals. In addition to the participating teachers and students, the CML provided all teachers and students in Grades 7–12 with the aforementioned software. The grade level range for participants in the third year was K–12. Year 3 of CML PD emphasized the use and production of media for storytelling. Participants were asked to explore how students can use mobiles to represent and convey their understanding of course content (e.g., Comic Life, Keynote, Photos, iMovie [Apple Inc., 2017b, 2017c]), Educreations (Educreations, Inc., 2017). Participating teachers were also required to create lesson plans that were compiled into an iBook using iBooks Author (Apple Inc., 2017a). The CML encouraged participants to continue in the professional development from year to year to further expand their repertoire of mobile learning.

The CML provided professional development through two primary methods. First, we provided intensive and ongoing support to cohorts of teachers in a year-long program. Each cohort participated in a multiday intensive Summer Institute that was typically 4 days long. Each cohort also received ongoing support and professional development through monthly meetings during the academic year. The academic year meetings focused on reinforcing and further elaborating upon concepts and strategies introduced during the Summer Institute. The meetings also allowed CML personnel to introduce new ideas and concepts and allowed the participating teachers to share and discuss the successes and challenges they faced in using mobiles in their classrooms. Challenges noted by participants focused on issues arising when students accessed the district wireless network at the same time, time required for software updates, their limited knowledge about certain applications, and district procedures for requesting and loading new applications on students' devices. Participants noted successes ranging from figuring out ways to easily share information to students using mobiles, tricks they learned about certain applications, and which applications their students were able to learn quickly and use effectively.

Through the Summer Institute and academic year meetings, the CML PD offered a variety of learning experiences and opportunities for teachers to examine technology, pedagogical strategies, and statewide initiatives for teaching and learning. For example, the CML explored and provided training for the software and approved apps made available in the participating teachers' district. They also developed profiles for the participants and organized them by grade level, content areas, and areas of interest for certain activities. This strategy allowed the participants to form small teacher teams in order to discuss and investigate topics and tools relevant to their needs and interests. Finally, we introduced, examined, and modeled inquiry-based teaching. Participants used the 5E instructional model (engage, explore, explain, elaborate, and evaluate) to support inquiry-based lesson planning (Bybee, 2015). The participants also were asked to design 5E inquiry-based lessons utilizing mobiles as part of the professional development experience. This approach provided three key benefits: (a) allowing us to support teachers across grade levels and disciplines, (b) enabling us to integrate mobiles and technological tools in the learning experience, and (c) permitting us to emphasize and integrate 21st century skills in the learning experience. To aid in the development of inquiry-based instruction through the use of mobiles, participants in the professional development program, as well as all the Grades 6–12 teachers and students in our partner district, were provided with software from the CML. This software included a concept-mapping tool (Inspiration) and a tool for digital storytelling (Comic Life). Although other software was available to participants, the overwhelming majority identified these two programs as essential to using mobiles in inquiry instruction. Teacher motivation alone was a reason to focus on these particular programs. In addition, Inspiration and Comic Life are inherently inquiry-oriented because they are geared toward acting as a tool for representing and communicating student thinking when examining different topics or problems.

Finally, the CML PD offered three supplemental workshops that enabled participants and other practicing teachers to engage in content specific support. The workshops also allowed collaborations across schools and encouraged participants to share and examine successful strategies and tools for integrating mobiles into their teaching. One supplemental workshop was a 2-day workshop for 6th–12th grade science teachers that supported teachers in using mobiles to share and visual data in real time.

EXAMINING OUR INTEGRATED MODEL OF PROFESSIONAL DEVELOPMENT

To examine the effectiveness of the CML PD in terms of encouraging innovative mobile use and high self-efficacy for integrating technology into their classrooms, we used a qualitative case study approach. Case study research is used to "investigate a contemporary phenomenon (the "case") in depth and within its real-world context, especially when the boundaries between phenomenon and context may not be clearly evident" (Yin, 2014, p. 6). Since this approach is effective in understanding "individual, group, organizational, social, political, and related phenomena" (Yin, 2014, p. 5), we were able to present the story of CML PD participants' self-efficacy for and use of mobiles for effective teaching. Using a holistic multiple case study approach (Yin, 2014) we present each participant as a separate case, and identify themes across cases to address the following research questions: (1) What was the nature of CML PD participants' self-efficacy for integrating mobiles throughout the CML PD? (2) What teaching practices did CML PD participants identify as being effective in integrating mobiles into their classroom practice?

Methods

Participants

Through purposeful sampling, we selected participants who were engaged in the Year 2 and Year 3 of the CML PD activities. Since Year 1 was used to develop a mobile learning profile of our participating school district and participation, extensive data collection began in Year 2 and continued in Year 3. We selected individuals who completed all CML PD required activities during those years (e.g., pre- and postsurveys, completed activities during and in between meetings, and attended meetings). We identified six participants, one middle school teacher, and five high school teachers, who met these criteria. As seen in Table 9.1, two participants taught science, one taught mathematics and engineering, two taught social studies, and one taught English language arts. Of this group, the middle school teacher was the only one who also participated in all three years of the CML PD. One participant's role at her school changed from teacher to administrator during Year 3. Three participants also attended the supplemental science workshop. These participants had intensive experience with content specific hardware and software, LabQuest 2 device, probes, and application. This content specific hardware and software allowed for data visualization, collection, analysis, and sharing both in real time and after the conclusion of a science investigation.

Data Collection

To develop an understanding of participants' views and use of mobiles and student-centered instruction, the following data was collected:

Table 9.1. Participants

Participant	*Grade Level*	*Content Area*	*Participation*
Chandra	High school	Science	Year 2, Year 3, supplemental science workshop
Darby	Middle school	English language arts	Year 1, Year 2, Year 3
Joanie	High school	Mathematics and engineering	Year 2, Year 3, supplemental science workshop
Kayla	High school	Science	Year 2, Year 3, supplementary science workshop
Molly	High school	Social studies (Year 2); Assistant principal (Year 3)	Year 2, Year 3
Nancy	High school	Social studies	Year 2, Year 3

(1) pre-, mid-, and post-TPACK surveys, (2) participant products created during CML PD meetings, and (3) classroom observations or interviews. Multiple data sources were used to address confirmability, or construct validity (Yin, 2014) and to allow for the convergence of evidence through data triangulation.

Participants completed a pre-, mid-, post-Technological and Pedagogical Content Knowledge (TPACK) survey (Koehler & Mishra, 2009; Mishra & Koehler, 2006) prior to and following professional development treatments. Reliability for this instrument was established during all three implementations. Each yielded high levels of reliability as determined through an analysis of the Cronbach's alpha (.948, .943, and 9.32 respectively). The TPACK uses a Likert-type scale with responses coded as follows: (5 = *strongly agree*, 4 = *agree*, 3 = *neither agree or disagree*, 2 = *disagree*, 1 = strongly disagree). Scores for the seven constructs of the TPACK instrument were summed across items to examine and compare descriptive statistics for pre-, mid-, and postscores. The constructs represented are as follows:

1. Technology Knowledge: TK (Items 1–6);
2. Content Knowledge: CK (mathematics, social studies, science, and literacy) (Items 7–18);
3. Pedagogical Knowledge: PK (19–25);
4. Pedagogical Content Knowledge: PCK (Items 26–29);
5. Technological Content Knowledge: TCK (Items 30–33);
6. Technological Pedagogical Knowledge: TPK (Items 34–42); and
7. Technology Pedagogy and Content Knowledge: TPACK (Items 43–46).

In addition to the Likert scale items noted above, we examined participants' responses to the open-ended TPACK item (Item 59). Item 59 states,

> Describe a specific episode where you effectively demonstrated or modeled combining content, technologies and teaching approaches in a classroom lesson. Please include in your description what content you taught, what technology you used, and what teaching approach(es) you implemented. If you have not had the opportunity to teach a lesson, please indicate that you have not.

We collected participants' responses to this question at three points over Year 2 and Year 3 (pre, mid, and post).

Throughout each phase of the CML PD, participants were asked to develop products to explore ways to best integrate inquiry and mobiles into their classroom. Participant products that we collected were Learning Process concept maps, inquiry-based lesson plans, 21st century skills presentations, content area skills outlines and reflective writings. During Year 2, participants were engaged in looking at definitions, processes, and evidence of learning. In doing that, they were asked to create concept maps to outline what learning is, what occurs during learning, and what is evidence of learning. Since inquiry-based instruction was used as the common thread for all subject areas, participants were asked to develop inquiry-based lessons that integrated mobile learning. Participants developed and shared their inquiry-based lessons plans in iBooks Author and Inspiration. Participants were also asked to create Keynote presentations to describe practices they use in their classroom to encourage students to engage in 21st century skills while integrating mobile learning. The last document collected was Content Area Skills outlines. To complete these outlines, participants in the same content area collaborated to identify (a) a content specific skill (e.g. balancing equations) that they would like students to work on, (b) how the content specific skill will be used and assessed in their classroom, and (c) how they can support the skill through inquiry-based instruction and mobile learning integration. As participants engaged in CML PD activities, they use Edmodo (2017) (Year 2) and Canvas (Instructure, Inc., 2017) (Year 3) to develop and submit reflective writings. The reflective writings prompted participants to examine how inquiry, 21st century skills and mobiles were implemented in recent lessons and which of lessons they believed are effective.

With the exception of one participant (Molly), the participants were observed twice during Year 3. Since Molly's role changed from classroom teacher to assistant principal at the beginning of Year 3, we were not able to conduct classroom observations of her practice. Instead, we conducted an interview to discuss the practices she has observed in her school.

For the other participants, their instruction was observed to see how inquiry, mobiles, and 21st century skills were being addressed. Participants selected the date, time, and lesson they wanted observed. During the observation, we noted what technologies were being implemented and who implemented and initiated each technology. Using a modified version of the Essential Features of Inquiry continuum (National Research Council, 2000) that was reworded to focus on multiple subject areas, we looked where the overall lesson fell on the continuum of inquiry. Using a checklist and definitions, we also identified which 21st century skills were being addressed and how the teacher attempted to encourage them in the classroom.

Data Analysis

We used a multiple case study approach (Yin, 2014) to examine each participant's self-efficacy for and experiences with mobile learning. The object of each case was one of the six participants. Open coding (Cortazzi, 2001; Polkinghorne, 1995) was used to identify themes in all qualitative data sources. Twenty-first century tools and their definitions (Partnership for 21st Century Learning, n.d.) were used as a priori content-specific themes (Schwandt, 2007) for coding the emerging themes for evidence of collaboration, communication, creativity, critical thinking, feedback, innovation, presentation, problem-solving, productivity, reflection, and social. After themes were categorized, categories were defined and revisited to attend to external and internal homogeneity (Patton, 2015). Cross-case analysis was used to examine similarities and differences among cases (Patton, 2015).

Due to the low sample size included in this study, we limited quantitative analysis of the TPACK to descriptive statistics. Inferential statistics were used to compare the overall dataset for the CML PD implementation over 3 years, but the case study design was used to look more deeply at a smaller sample of this larger dataset rather than make generalizations about the sample as a whole. We examined summed and mean scores for each TPACK construct to gain an overall understanding of shifts that occurred from pre to mid to post implementations.

Findings

The findings represent the cross-case analysis conducted for the six participants' cases and is organized according to the research questions.

Nature of CML Participants' Self-Efficacy for Integrating

Table 9.2 outlines the mean construct scores from pre to mid to post for each participant. While the seven TPACK constructs provide an overall picture of what teachers need to know to use technology effectively, we

Table 9.2. Mean Construct Scores

		Chandra	*Darby*	*Joanie*	*Molly*	*Nancy*	*Kayla*
Pedagogical Content Knowledge	Pre 2014	3.00	3.50	4.75	5.00	4.50	5.00
	Mid 2015	4.00	3.75	4.00	4.50	4.00	4.25
	Post 2016	4.25	4.25	4.75	4.00	4.00	3.75
Technological Content Knowledge	Pre 2014	3.25	2.75	5.00	4.75	3.50	2.50
	Mid 2015	3.75	3.75	4.25	4.50	4.00	2.75
	Post 2016	3.25	3.50	5.00	4.50	4.00	3.25
Technological Pedagogical Knowledge	Pre 2014	4.11	3.78	5.00	5.00	4.44	4.11
	Mid 2015	4.44	4.33	5.00	4.78	4.89	4.33
	Post 2016	4.33	4.89	5.00	5.00	4.00	4.00
Technology Pedagogy and Content Knowledge	Pre 2014	4.00	2.75	4.50	4.75	4.00	2.75
	Mid 2015	3.75	3.75	3.75	4.25	4.50	4.25
	Post 2016	4.25	3.50	5.00	4.50	4.00	3.25

focused analyses on four of the constructs (Pedagogical Content Knowledge, Technological Content Knowledge, Technological Pedagogical Knowledge, and Technology Pedagogy and Content Knowledge). The rationale behind this choice lies in the theoretical framework for the TPACK, that the first three constructs (Technology, Content, and Pedagogy) represent the three areas of knowledge needed to understand how to use technology, but the remaining four constructs allow us to see the connections between these three areas of knowledge.

The participants all identified themselves as having a moderate to high self-efficacy for pedagogical content knowledge related to their subject area expertise. With the exception of two participants, Molly and Nancy, participants' scores from pre- to post-TPACK for Technology Pedagogy and Content Knowledge increased. Molly's self-efficacy was noted as decreasing by 0.50 by the post-TPACK while Nancy's remained the same.

This decreasing trend from pre to post for Molly also occurred for two other constructs (Technological Content Knowledge and Pedagogical Content Knowledge). However, Molly's mean scores overall indicated she had a higher level of technology self-efficacy than the other participants in this sample. Overall, the strongest shifts occurred from pre to post in the Technological Pedagogical Knowledge construct. This finding is particularly poignant because we had multiple content disciplines represented in this sample. The focus of this mobile learning professional development was really geared toward the technological pedagogical knowledge construct because we intentionally chose a cross-disciplinary approach to the professional development

Effective Mobile Learning Integration. As participants identified or discussed effective practices for integrating technology in TPACK Item 59, content area skills outlines, and their 5E lesson plans, the following themes emerged: (a) integrating mobiles to provide students with opportunities to communicate in creative ways, (b) integrating applications to allow students to explore real-world applications of content, and (c) integrating applications to provide students with instant feedback and autonomous exploration of content. These three themes were reminiscent of learning evidence that was noted on CML PD participants' learning process concept maps. In Chandra's learning process concept map, she identified performance-based tasks, ability to explain procedures, ability to use vocabulary and ability to help peers as evidence that students have learned a concept. Like Chandra, the other participants' illustrated activities where students completed traditional and performance assessments and explained content using correct vocabulary as evidence of learning. It is obvious that these ideals were embedded in participants' understanding of effective teaching in that these assessments were used as participants developed 5E lessons and identified effective lessons they taught.

Creatively communicating content and developing explanations was a popular use of mobiles in the classroom. Mobiles applications, such as using "iPhoto, iMovie, Keynote, Comic Life ... to explain a topic they had [sic] studied" (Kayla, TPACK Item 59), supported students in communicating their understanding to teachers and their peers. For example, Chandra noted,

> I asked students to use ComicLife or other applications to teach others how to solve a problem. They were given choices from many topics we had covered. They had to give an example and show the steps they chose to describe. Most students did use Comic Life just because it is well designed to do this quickly and I used it as a model. Others used iMovie or Keynote because they enjoyed the creativity or felt more comfortable with the application (TPACK Item 59).

Some of Chandra's students used ComicLife to describe how to draw electron dot diagrams to illustrate the bonding between atoms of a molecule. Similarly, other CML PD participants also used applications that encouraged digital storytelling to showcase their understanding of specific content. The use of evidence to support explanations is a result of creatively explaining content. Participants noted that students needed to think critically about content and use evidence, not opinions, to show that they truly understood content. "The use of technology helped with this aspect, because they were able to conduct research and gather supporting evidence displaying using Photos, Keynote, or iMovie" (Nancy, Reflective Writing 1). Student-generated products that emphasized creativity and afforded students to opportunities to easily share a variety of evidence encouraged 21st century tools of communication, creativity, and critical thinking (P21, n.d.).

Mobile integration that explored real-world connections to classroom topics made content relevant and "students form a connection to the content" (Nancy, Reflective Writing 1). Web-based applications, such as Google Earth (Alphabet Inc., 2017), that present real-world data and web searches that focused on real life applications of content were often seen as important classroom activities. In addition to asynchronous applications, some teachers also used synchronous applications to collect information from sources other than websites. In one of her lesson on South East Asia, Molly and her students "Skyped with a teacher in India to better understand Indian culture. [They] also read headline news articles from Indian newspapers and compared them to headline news articles from America" (TPACK Item 59).

Using mobiles to access online simulations and games that present instant feedback and autonomous learning opportunities were commonly used to supplement classroom activities or present information during class. Joanie, for example, used "http://illuminations.nctm.org to model linear and quadratic functions in an inquiry-based lesson [so that] students could actively see how changing coefficients, y-intercepts, etc. would change the lines or parabolas in real time" (Joanie, TPACK Item 59).

Participants also identified the need to use mobiles for encouraging student-driven or autonomous learning of content. In a discussion of one lesson, Joanie noted,

> I used Educreations, LearnZillion (2017), Khan Academy and many other online resources for content. One lesson in particular introduced how to write an equation of a line when given the slope and y-intercept. I used Educreations to introduce the lesson using guided notes. Learn Zillion followed with a video lesson and practice. After this introduction phase I followed up with 4 stations that allowed students to participate in more independent practice (TPACK Item 59).

In this example, students explored content through online resources prior to engaging in classroom activities that reinforce the content. Since we do not have classroom observation notes to indicate the amount of inquiry that was present in the lesson, we can identify that Joanie attempted to present content to students for autonomous exploration via web-based applications. Like Joanie, the other participants also used mobiles to encourage students' information literacy skills through accessing and managing content.

Issues With Implementation of Mobile Learning

While CML PD was able to provide a wealth of ideas for mobiles that were both readily accessible to participants at their schools as well as provide additional apps and resources, there was dissonance with the school district structure and capabilities. Three main issues occurred during implementation that may be common to conducting professional development with mobile learning. The first issue was that it took longer than expected for students to receive software, which resulted in a large part to the massive implementation of iPads for Grades 2–5 and MacBooks for Grades 6–12 for the entire school district. The additional wait time was both a benefit and constraint to participants. It was beneficial in that teachers had additional time to familiarize themselves with the software and use it in "teacher-driven" mode before the students received the software. It was a constraint in that teachers were not able to immediately integrate certain applications due to them not being accessible to students. The second technological issue that was necessary to troubleshoot was the limited accessibility of certain websites at some school sites. While it understandable that protocols have to be in place to prevent unauthorized and inappropriate access, it can be crippling if web sites and/or web-based tools are inaccessible due to network restrictions. The final issue was that the school district did not have the infrastructure in place to handle the bandwidth necessary for supporting the use of mobiles by teachers and students across the district. This limitation had a significant impact on how and when teachers began integrating certain learning-focused activities.

These limitations can be common when working with school districts as they implement mobile learning on a large scale. It is important to be aware of these issues prior to designing professional development and providing mobile and applications to teachers and schools. Our approach to addressing these issues was to identify the accessible apps, websites, and web-based tools that would be commonplace and instantly accessible on the mobile devices. We started off by providing support for these technologies and incorporated others identified as useful by participants in the professional development experience. As issues with network and infrastructure were reduced, we began to model, provide, and engage participants in exploring additional mobile technologies. We addressed the lack of mobile traffic and time to fully implement mobiles by revising timeline expectations for teachers to submit artifacts and discussing ways to troubleshoot these issues during professional development sessions.

MAKING A CASE FOR INTEGRATING MOBILE LEARNING WITH AN INQUIRY-BASED APPROACH

The growth and adoption of mobiles in K12 classrooms is going to become more common in the years to come. Because of this shift, effective professional development models for integrating mobile learning on a small or large scale are critical to adoption success. This study illustrates that the use of mobiles, whether it be laptops or smaller screened, handheld devices, coupled with inquiry-based teaching is an effective approach to provide several affordances to teachers and students. Our participants maintained a high self-efficacy for integrating technology into their content area instruction and integrated applications to give students' ownership of their learning experiences. While we can only attend to analytical generalization for this qualitative study, findings give us insight into how dual emphases on mobiles and inquiry can encourage creative uses of mobiles, student-generated products, and 21st century tools.

Using an inquiry-based framework, we supported teachers across multiple grade levels and disciplines in a single professional development experience. Participants collaborated within and across disciplines to integrate mobiles in inquiry-based instruction to foster student creativity and ownership of learning. Without the use of mobiles, their students could not access, collect, and share content throughout all of their inquiry activities. The mobility and flexibility of technologies selected by CML PD participants were critical to creating an environment where the students could move from independent to collaborative situations within seconds while participating in a variety of learning activities. Mobiles allowed students to move throughout their learning environments to easily collect and analyze data in real time. Even though less portable technologies allow students to record, analyze, and share data, they do not allow students to quickly reorganize into groups, move among laboratory stations, or spontaneously collect pictures to use for digital storytelling. Through applications providing instant feedback, students were presented with opportunities for in-the-moment reflection on their abilities to effectively use mobiles, as well as their content knowledge understanding. Students were able to use software in any location within or outside of their classroom environment to communicate and complete assignments. Mobiles allowed teachers to adapt their current learning configurations and create learning spaces on demand. Teachers could further cultivate learning communities in their classroom by quickly organizing students into groups during collaborative learning activities. Thus, the technology adapted to the learning environment as opposed to the learning environment having to adapt to the technology.

Another key finding from this study was that situating and aligning the professional development to classroom needs and constraints was beneficial in that it made the experience more meaningful to the participating teachers. Our professional development model allowed facilitators to integrate key concepts and themes related to 21st century learning such as communication, collaboration, creativity, and critical thinking while supporting mobile learning. Through needs and interest assessments, we identified tools that can be used across contexts (i.e. grade levels and content areas) that enabled the participants to think more broadly and collaboratively about products students can create (e.g., video, audio, comics, etc.) across disciplines rather than focusing on subject specific applications. This approach aligns with Hoyle's (1982) vision of professional development in that it enabled the participating teachers to design lessons and units that were relevant to their needs and the needs of the students. Lastly, this approach also allowed the facilitators to build a community of learners through a network of groups. Participants collaborated with content area peers to identify and adapt approaches for using mobile devices and tools for their specific classrooms. These practices are reminiscent of characteristics of effective professional development encouraged by Loucks-Horsley et al. (2010) and Posnanski (2002).

CONCLUSION AND IMPLICATIONS

This study demonstrated that providing professional development using an inquiry-based framework for mobile learning was an effective strategy for teachers to identify ways to integrate mobiles and mobile tools into their classroom. This approach allowed the PD facilitators to work with participants across grade levels and content areas. For example, we could focus on a specific part of a 5E lesson as the context for integrating specific applications or approaches to content (e.g., media production). Rather than being tied to a content-specific tool, the PD facilitators were able to empower the participants to identify and adopt the tool to fit their own curricular needs or learning goals for their students. This approach enabled the professional development facilitators to make the experience authentic and relevant to the needs of the participating teachers while having a common context as the thread. The facilitators were also aided by having a common platform that was available to all participating teachers (Mac OS or iOS). This commonality allowed the facilitators to spend a majority of their time on the tools and their application rather than software training across multiple platforms. This commonality also benefited teachers because they had knowledge and confidence that the tools being used would work universally for their students as well. Facilitators could be flexible and malleable in how they approached supporting teachers in using mobiles because they were not tied solely to the devices themselves. They were able to shift approaches if needed due to technical limitations, challenges (e.g., bandwidth or network restrictions), or content area concerns. We recommend using an inquiry-based framework to introduce the use of mobile devices in the classroom because it provides several key benefits to students. Those include providing students with new line of communicating information in creative ways, being able to explore content authentically, and enabling autonomous exploration of content.

ACKNOWLEDGMENT

The activity, which is the subject of this chapter, was produced under a grant from the South Carolina Commission on Higher Education under the auspices of the EIA Teacher Education Centers of Excellence Grant Program.

REFERENCES

Alphabet Inc. (2017). Google Earth for iOS [Software]. Available from https://appsto.re/us/rffGr.i

Anderson, J., Reder, L., & Simon, H. (1996). Situated learning in education. *Educational Researcher, 25*(4), 5–11.

Apple Inc. (2017a). iBooks Author [Software]. Available from http://www.apple.com/ibooks-author/

Apple Inc. (2017b). iMovie [Software]. Available from https://appsto.re/us/r0rFw.i

Apple Inc. (2017c). Keynote [Software]. Available from https://appsto.re/us/ODmIv.i

Apple Inc. (2017d). Photos for iOS [Software]. Available from http://www.apple.com/ios/photos/

Brown, J., Collins, A., & Duguid, P. (1989). Situated cognition and the culture of learning. *Educational Researcher, 18*(1), 32–42.

Bybee, R. (2015). *The BSCS 5E instructional model: Creating teachable moments*. Arlington, VA: NSTA Press.

Cortazzi, M. (2001). Narrative analysis in ethnography. In P. Atkinson (Ed.), *Handbook of ethnography* (pp. 384–394). London, England: SAGE.

Crompton, H. (2013). Mobile learning: New approach, new theory. In L. Z. Berge & L. Y. Muilenburg (Eds.), *Handbook of mobile learning* (pp. 47–58). New York, NY: Routledge.

Edmodo. (2017) Edmodo [Software]. Available from https://www.edmodo.com

Educreations, Inc. (2017). Educreations Interactive Whiteboard [Software]. Available from https://appsto.re/us/L3XHC.i

Guskey, T. (1995). *Results-oriented professional development: In search of an optimal mix of effective practices.* Arlington, VA: North Central Regional Educational Laboratory.

Hedegaard, M. (1998). Situated learning and cognition: Theoretical learning and cognition. *Mind, Culture, and Activity, 5*(2), 114–126.

Hoyle, E. (1982). The professionalization of teachers: A paradox. *British Journal of Educational Studies, 30*(2), 161–171.

Inspiration Software, Inc. (2017). Inspiration Maps [Software]. Available from https://appsto.re/us/8BNzE.i

Instructure, Inc. (2017). Canvas [Software]. Available from https://www.instructure.com

Khan Academy, Inc. (2017). Khan Academy [Software]. Available from https://appsto.re/us/z0yaC.i

Kearney, M., Schuck, S., Burden, K., & Aubusson, P. (2012). Viewing mobile learning from a pedagogical perspective. *Research in Learning Technology 2012, 20,* 14406. doi:10.3402/rlt.v20i0/14406

Koehler M. J. & Mishra, P. (2009). What is technological pedagogical content knowledge? *Contemporary Issues in Technology and Teacher Education, 9*(1), 60–70.

Kukulska-Hulme, A. (2010). Mobile learning as a catalyst for change. *Open Learning: The Journal of Open and Distance Learning, 25*(3), 181–185.

LearnZillion. (2017). LearnZillion [Software]. Available from https://appsto.re/us/nc2l6

Loucks-Horsley, S., Stiles, K. E., Mundry, S., Love, N., & Hewson, P. W. (2010). *Designing professional development for teachers of science and mathematics* (3rd ed.). Thousand Oaks, CA: Corwin Press.

Mifsud, L. (2014). Mobile learning and the socio-materiality of classroom practices. *Learning, Media, and Technology 39*(1), 142–149.

Mishra, P., & Koehler, M. (2006). Technological pedagogical content knowledge: A framework for teacher knowledge. *Teachers College Record, 108*(6), 1017–1054.

National Council of Teachers of Mathematics. (2017). Illuminations [Software]. Available from https://illuminations.nctm.org/

National Research Council. (2000). *Inquiry and the National Science Education Standards: A guide for teaching and learning*. Washington, DC: The National Academies Press.

Ozdamil, F. (2012). Pedagogical framework of m-learning. *Social and Behavioral Sciences, 31*, 927–931.

Partnership for 21st Century Learning. (n.d.). P21 framework definitions. Retrieved from http://www.p21.org/about-us/p21-framework

Patton, M. (2015). *Qualitative research and evaluation methods* (4th ed.). Thousand Oaks, CA: SAGE.

Plasq, LLC. (2017). Comic Life 3 [Software]. Available from https://appsto.re/us/iMvi1.i

Polkinghorn, D. (1995). Narrative configuration in qualitative analysis. *Qualitative Studies in Education, 8*(1), 5–23.

Posnanski, T. J. (2002). Professional development programs for elementary science teachers: An analysis of teacher self-efficacy beliefs as a professional development model. *Journal of Science Teacher Education, 13*(3), 189–220.

Schwandt, T. (2007). *The Sage dictionary of qualitative inquiry* (3rd ed). Thousand Oaks, CA: SAGE.

Yin, R. K. (2014). *Case study research: Design and methods* (5th ed.). Thousand Oaks, CA: SAGE.

CHAPTER 10

DIGITAL LEADERSHIP IN RURAL MIDDLE SCHOOL

Preparing Effective 21st Century Digital Citizens Via a Mobile Learning Curriculum

Jennifer L. Motter
Forest Hills Junior-Senior High School, Sidman, PA

In a digital society, pre-K–12 schools cannot afford to ignore contemporary technologies' capabilities. Mobile devices should be smoothly *interwoven*, rather than sparsely deployed, across the curriculum, enabling them to function as familiar instructional tools necessary for daily learning within educational settings (Saltman, 2014). Mobile devices influence our daily lives, teaching, and learning by offering rich resources, including multimedia content, anytime and anywhere (Sheninger, 2014). New media, such as the Internet, online social networking, virtual worlds and games, blogs, and wikis, is identified by the following characteristics: digital, interactive, hypertextual, virtual, networked, and simulated (Lister, Dovey, Giddings, Grant, & Kelly, 2009).

New media enables message receivers, or consumers of content, to transform into content creators who contribute to virtual world discourse by sending personal messages to a public audience via the Internet. As

Mobile Learning: Perspectives on Practice and Policy
pp. 211–229

educators, we must guide content creators in the classroom. Providing information and communication technology for teaching and learning is no longer enough, educators must instill students with the capacity and confidence to design digital technologies to solve real-world issues (Nair, 2016). Mobile learning facilitation enables educators to promote productive "play" by encouraging students to creatively contribute to virtual world discourse, rather than function as passive information consumers through skill-and-drill testing (McMunn Dooley, Lewis Ellison, Welch, Allen, & Bauer, 2016).

In today's world, communication via new media enables people of diverse backgrounds to enter the conversation by creating public forums in virtual space providing opportunities for a plethora of actions, voices, and interactions. Mobile devices, including iPads, tablets, laptops, and smartphones, enable us to conveniently engage in new media discourse anytime and anywhere, as long as we have a cellular signal and Wi-Fi. Mobile technologies can be leveraged to provide users with voice and choice in and out of the classroom (Scheffer, 2016).

Thus, educators must properly prepare students for our "participatory culture" (Jenkins, Purushotma, Clinton, Weigel, & Robinson, 2007) in which many individuals, including youth, engage in new media, such as online social networks, using mobile technologies. Mobile devices support our participatory culture, which offers many opportunities for artistic expression and civic engagement, strong support systems for creation and sharing, and informal mentorship that involves those with higher levels of experience passing on their knowledge to novices (Jenkins et al., 2007). Mobile learning in pre-K–12 today is significant as our "participatory culture is also one in which members believe their contributions matter, and feel some degree of social connection with one another" (Jenkins et al., 2007, p. 3).

In the 21st century, real and virtual are difficult to separate. According to McLuhan's deterministic view, media extends human's capabilities, saturates our environment, arouses our senses, and impacts culture and society (Lister et al., 2009). Social media, online games, virtual worlds, and mobile devices are now ingrained in youth culture (Ito et al., 2008; Liao, 2008). If students have access to mobile technologies, they are likely using it for various purposes, such as communicating, collaborating, networking, problem solving, informal learning, and creatively expressing themselves in their daily lives. In 2016, 86% of young people ages 18–29 used social media (Pew Research Center, 2017b). According to Pew Research Center (2017b), 88% of young people use the Facebook social media platform, 59% use Instagram, 36% use Pinterest and Twitter, and 34% use LinkedIn. Mobile devices, such as smartphones and iPads, enable us to view and share multimedia anywhere there is cellular service and a Wi-Fi connection. For example, the Instagram app allows users to take photos and videos, edit photos and videos by cropping and applying filters, and share them with an audience of followers from their present location.

It is educators' responsibility to not only keep students up-to-date with the latest technological innovations, but also promote ethical, collaborative, and critical uses of current technologies for positive productive and possibly empowering and transformative learning purposes (Julian, 1997; Keifer-Boyd, Amburgy, & Knight, 2007). Using the same digital tools in class that students use to explore personal interests outside of the classroom can spark students' interest and further engage them in classroom learning activities. To function effectively in society, students must possess new media knowledge and skills attained through mobile learning experiences using contemporary mobile technologies, as every educated citizen should possess digital knowledge and skills enhanced through educational institutes and workplaces (Léger & Freiman, 2016). Léger and Freiman (2016) found that adults entering the workforce who previously experienced a one-to-one laptop initiative in middle school possess the following digital skills that lead to higher education and career success: technological resourcefulness, digital self-efficacy or empowerment, and open-mindedness toward technology.

BRIDGING THE DIGITAL GAP: PROVIDING MOBILE TECHNOLOGIES TO RURAL MIDDLE SCHOOL STUDENTS

Current technology, new media and mobile devices, access, and usage experience in pre-K–12 schools are necessary for 21st century students (Dolan, 2016). In the past, the concept of the digital divide represented a personal computer and an Internet access equity issue in which a dominant population, consisting of White upper class males living in suburban areas, was primary personal computers users. Historically underrepresented groups including Blacks, Hispanics, women, lower socioeconomic class, and those living in urban and rural areas had significantly less opportunities to access and experience computers and the Internet. Thus, minority groups were left behind in terms of technological advancement (Picciano, 2011). Other factors that influence technology experience include parental influences, environmental influences, gender stereotypes, language barriers, time, and skills (Hughes, Read, Jones, & Mahometa, 2015).

Today, the digital divide gap is narrowing because of minority groups attaining more education and income (Picciano, 2011). However, it still exists, particularly for those who are less educated, lower income, older, or living in rural areas of America where there is limited mobile technology options and a lack of high-speed Internet connectivity (Wheeler, 2014; White House, 2015). Only 81% of people living in rural communities use the Internet, while 89% of people living in urban communities and 90% of people living in suburban communities are Internet users (Pew Research Center, 2017a). President Obama confirmed that "Today high speed broadband is not a luxury, it's a necessity" (White House, 2015).

In many educational institutes outside of rural areas, the digital divide is now largely based on the quality of educational technology students experience, not Internet access (Sheninger, 2014). Smartphones have helped bridge the divide. More people around the world, including previously excluded minority groups, have access to personal technologies and the Internet. Minority groups such as Blacks, Hispanics, girls, and women use smartphones equally or more than White boys and men (Hall, 2013). pre-K–12 students comprise a large percentage of mobile technology users. According to Dolan (2016), 56% of middle school students in Grades 6–8 and 68% of high school students in Grades 9–12 have Internet access in their hands via their mobile devices.

Unfortunately, the digital divide gap is even greater for those attending rural school districts where students are behind in digital literacy practices (McMunn Dooley et al., 2016), due to no or slow broadband connectivity and cellular service at school or home. Upgrading Internet and phone services is not prioritized in low populated areas; therefore, those living in some rural communities are isolated from the rest of the digital world (Thielman, 2016). Some rural schools are still struggling to obtain high-speed connectivity, which causes rural students to fall behind their privileged urban and suburban peers (Herold, 2015). At Flood Valley, a rural school district in Western Pennsylvania, the lingering existence of a digital divide exists, as some students still do not have Internet access at home, or personal mobile devices. There will always be a technology affordability gap between the rich and the poor. While personal technologies, such as mobile devices, are becoming less expensive, upper socioeconomic groups will likely possess the newest innovations first. Later, these technologies will become affordable for others, but the elite will always have an upper hand (Hall, 2013). However, it is the job of educators to ensure that all students receive the same chance to engage in mobile learning as digital participants, regardless of their socioeconomic status (McMunn Dooley et al., 2016). Therefore, to prepare rural students to be effective citizens in the 21st century, we must provide them with new media and mobile technology experience, knowledge, and skills in the rural classroom.

EMBRACING PROGRESSIVE MOBILE TECHNOLOGIES IN PRE-K–12

Regardless of their locale, pre-K–12 schools should strive to keep current with technological advancements, including new media and mobile technologies. With proper planning, mobile devices and one-to-one initiatives have great potential to benefit underserved and remote regions, as mobile devices are more affordable and easier to widely distribute and deploy due to their lightweight portable nature, require less infrastructure and electricity than desktop computers as wireless connectivity becomes more widespread, and can provide increased access to learning opportunities (Harper & Milman, 2016; Kim et al., 2011). In the past, and continuing today, planning for technology in schools is inadequate. According to Picciano (2011), technology in schools progressed from print to electronic media in the later 20th century. However, schools today are still behind current business practices, which presents problems for students entering a 21st century workforce as they may lack the knowledge and skills necessary to compete and strive. Time and budget are the main reasons for the lack of effective technology implementation in schools.

Given the need to expose youth to new media and mobile technologies in a rural educational setting in order to provide positive mobile learning experiences and encourage safe, responsible, and productive engagement with 21st century digital tools, the author, Jennifer Motter, presents a mobile learning curriculum designed to facilitate student development of 21st century knowledge and skills. Through the examples, Motter aims to offer insight into mobile learning curriculum development and implementation at a rural middle school and encourage other educators' implementation of mobile technologies in pre-K–12.

MOBILE TECHNOLOGY COURSE DEVELOPMENT

The course, *Mobile Technology*, was developed during Motter's principal internship in spring 2015 to provide rural middle school students with hands-on learning experiences using mobile devices. Prior to executing the course, administrators and teachers carefully considered logistics including associated costs, mobile device selection, teacher assignment, tech team support, and student scheduling. *Mobile Technology* placed mobile devices in the hands of each eighth-grade student in the hope of contributing toward significant efforts to eliminate the digital divide that still exists within Flood Valley and that impacts other students attending rural school districts (Wheeler, 2014; White House, 2015). Providing students with access to different devices is particularly important for those who don't have mobile devices at home. By providing middle school students with an opportunity to experience different mobile devices and digital learning tools, the *Mobile Technology* course objective was to effectively prepare 21st century middle school students to be responsible digital citizens who are productive and successful in their current and future digital lives, including college and the workplace, where technology literacy will be assumed, while broadening their horizons (Butrymowicz, 2014). The *Mobile Technology* course description reads:

> Students will obtain 21st century knowledge and skills through project-based learning via mobile devices. During the semester, students will have hands-on experience using different mobile devices and various apps while engaging in International Society for Technology in Education Standards-aligned curriculum covering creativity and innovation; communication and collaboration; research and information fluency; critical thinking, problem solving, and decision making; digital citizenship; and technology operations and concepts. Topics explored include: mobile photography and video, mobile creation, design, and development, mobile 3-D modeling, mobile presentations, mobile storage and sharing, mobile writing, mobile productivity, and mobile subject-specific apps.

Curriculum Mapping for Mobile Learning

Conceiving, planning, and preparing for a previously nonexistent course can be challenging. However, many helpful online resources exist including ISTE Standards for Students (International Society for Technology in Education, 2007), GCF LearnFree Windows 10 tutorial (Goodwill Community Foundation, 2016), and Common Sense Media's (2016b) digital citizenship unit plans. G Suite for Education accounts and apps (Google, 2016) are extremely effective tools for coplanning that enable asynchronous collaboration, communication, and resource sharing with ease, despite hectic teaching schedules. For example, a Google Sheet can be created and shared amongst colleagues for curriculum mapping of learning goals and objectives, standards, projects, and devices.

Mobile Learning Curriculum: Devices and Content

We began the 2015–2016 school year using Chromebooks and Lenovo Yoga Tablets, which were selected due to their portability and price (Sheninger, 2014). First, student Google accounts were created and setup. *Mobile Technology* was the first course at Flood Valley Middle School to use Google Classroom as a daily digital instructional tool and course management system. Therefore, students needed time to develop comfort and confidence using the Google Classroom app early in the school year. Throughout the school year, *Mobile Technology* teachers posted, collected, graded, and provided instructions and feedback on assignments; shared resources; and engaged students in online peer critique and reflection via the Classroom Stream. Next, students were introduced to G Suite for Education primarily including Classroom, Drive, Docs, Gmail, Forms, Sheets, and Slides. Then, students shifted to Lenovo Yoga Tablets.

After covering the G Suite for Education basics on the Chromebooks and familiarizing middle school students with the Windows 10 operating system on tablets to ensure they were comfortable using Lenovo Yoga Tablets, we focused on digital literacy and citizenship, including topics such as: online communication, Internet safety, privacy and security, copyright, cyberbullying, digital footprints, self-representation and online identity, digital information literacy, and online scams. For example, during the copyright unit students learned about image and text usage rights by watching *Copyright and Fair Use Animation* (Common Sense Media, 2016a), using Google Advanced Search and Google Advanced Image Search, and collaboratively working in small groups using Google Docs to create a list of debate points regarding appropriate image use for a fictional ad. As a result of studying copyright laws and requiring application of their knowledge in class, students were more mindful of plagiarism and cautious when selecting images to use for class presentations and research papers. Through the copyright unit, students learned appropriate fair use of copyrighted materials in educational settings. "Fair use is flexible, but it requires reasoning and interpretation of situational factors" (Hokanson, 2015, p. 11).

iPads were introduced to *Mobile Technology* students in spring 2016, a semester dedicated to creativity and innovation. iPads, the dominating tablet on the market, were selected for student use due to their wide and diverse app availably. iPads supported over 700,000 apps in 2013 (Sheninger, 2014). As experienced in *Mobile Technology*, iPad-based teaching and learning: facilitates differentiation and visualization of content and concepts; fosters independence and agency for students and teachers; and, is dynamic, fun, engaging, collaborative, and interactive (Smith & Santori, 2015). iPads were used for student group projects involving 21st century skills: collaboration, problem solving, critical thinking, communication, creativity, and innovation (Partnership for 21st Century Skills, 2007). The following creative projects were assigned to *Mobile Technology* students: Stop Motion Animation, iMovie Trailer, and Educational App Invention. These projects provided students with authentic learning experiences with mobile devices that connected to real-life experiences, motivated student engagement in learning, and prepared students for life outside of school (Hughes et al., 2015; Sheninger, 2014). iPad apps, such as Keynote (Apple Inc., 2016), Prezi Viewer (Prezi Inc., 2017), and SlideShark (Brainshark Inc., 2016), that afford cloud-based storing, sharing, and presenting are extremely valuable and useful in educational settings, as they provide many rich and innovative ways to engage students in creating, saving, and presenting course content.

IMPLICATIONS FOR MOBILE TECHNOLOGY IMPLEMENTATION

Today, as in the past, many schools still focus on bringing computers and other digital and mobile devices into classrooms, rather than considering their effective implementation, including who will use them, and how they will extend the curricula and enhance learning. A top-down approach in schools is still common, which involves administration insisting teachers use technology a specific way, for administrative or instructional tasks, rather than providing time for teachers to explore technology and its potential as a teaching and learning tool in the classroom. Teachers typically have little say regarding implementation of technology in their classes. Therefore, technology use tends to be limited and lacks innovation (Picciano, 2011). Motter is fortunate to have her input sought before technology decisions are made for her classroom.

With the task of effectively preparing digital citizens in mind, Motter and her colleague at Flood Valley selected the devices and apps to explore in *Mobile Technology*. Using apps such as Google Classroom, enabled *Mobile Technology* to be a paperless course. *Mobile Technology* teachers used Google Classroom to deliver digital assignment and project instructions, provide online resources and feedback, and facilitate online grading. Digital productivity tools such as mobile storage and writing apps have the potential to encourage progressive change. For example, using mobile storage apps would help cut school districts' printing costs, as many storage apps are free to download and offer sharing capabilities. Distributing and submitting hardcopies of assignments and purchasing text hardcopies is no longer necessary. Using school Google accounts, students signed in to Google Classroom each class to participate in *Mobile Technology* virtually by reviewing their online assignments, engaging in online discussions, and submitting their finished work online.

Pre-K–12 mobile device implementation affords learning across the curriculum. As a *Mobile Technology* instructor, Motter facilitated learning experiences involving collaborative creative work that engaged students in mobile cross-curricular learning. For example, after reading the article "Don't know the difference between emoji and emoticons? Let me explain" (Hern, 2015) and watching the YouTube video *Emoticons Are*

Changing Our Brains! (DNews, 2014) students worked in small groups to develop online surveys using Google Forms to distribute to their peers via e-mail in order to better understand their use of and attitudes toward emojis and emoticons. Small groups received peer online survey results in Google Sheets, and then collaboratively created a Google Slides presentation including graphs created in Google Sheets, "free to use share or modify" images found via Google Advanced Image Search, and text, as well as innovative emojis created using digital drawing tools in Google Drawings. These Google apps allowed students to each act as online editors of their collaborative work by individually contributing to their group work synchronously from their own devices during class time, or asynchronously contribute outside of class. In Google Docs, students utilized the chat feature and inserted digital comments while remotely discussing their group projects. As the course instructor, Motter also acted as the moderator requiring all group Docs, Forms, and Slides be shared with her, so she could see students' digital conduct and assignment participation. Mobile learning courses, such as *Mobile Technology*, assist students in attaining contemporary technology knowledge and skills they can apply in the future, as well as in their current core courses for major projects such as research papers and group presentations.

Mobile 3D Design and Printing

Mobile device implementation also offers authentic learning experiences to pre-K–12 students, and unique contemporary design and making opportunities. At Flood Valley, *Mobile Technology* students engaged in 3D printing using mobile devices. The process of 3D printing involves melting PLA, a biodegradable thermoplastic, to create an object constructed out of layers of plastic. Students created innovative 3D printing projects using Ultimaker and Ultimaker 2 Go 3D printers for individual and collaborative works. For example, students worked in small groups to create articulated bodies using the iPad Tinkerplay app (Autodesk Inc., 2016). Each group member digitally added at least one body part, to form a collaborative design. Tinkerplay also allows the user to add a digital environment and take a photo of his or her 3D body in its environment. Articulated bodies were downloaded for 3D printing, and later used in stop motion animation group projects created with iPad apps OSnap! (Cegnar, 2016) and Stop Motion Studio (Cateater, 2016).

During the 3D printing unit, online resources Thingiverse (http://www.thingiverse.com/) and Tinkercad (https://www.tinkercad.com/) were also explored by middle school students for digital 3D model design. Thingiverse allows the user to download STL model files, which students can then edit using the 3D design program Tinkercad. Using Tinkercad, students were able to design their own 3D objects after following user-friendly tutorials. This case demonstrates that mobile technology integration enables students to take on leadership roles, such as prototype designers, while developing 21st century knowledge and skills.

SUPPORTING STUDENT LEARNING: POTENTIAL OF MOBILE TECHNOLOGIES

The *Mobile Technology* course demonstrates how new media and mobile devices can facilitate, enhance, and further learning when appropriately used as instructional tools in the classroom. Motter found that mobile devices provide students with a plethora of easily accessible resources at their fingertips, and new media, such as G Suite for Education, extends opportunities for collaboration, creativity, communication, and informal learning. *Mobile Technology* students experienced learning with mobile devices that is convenient and interactive. As this case presents, exploring new media via mobile technologies affords innovative pedagogical potentials for fostering asynchronous and synchronous group work, as well as global peer-to-peer collaborative learning. The *Mobile Technology* course provides an example of a mobile learning course that supports self-directed life-long learning through integration of new media in union with mobile technologies.

Student perspectives provide evidence of *Mobile Technology*'s significance in potentially supporting life-long learning. As an instructor, Motter sought student feedback at the beginning, middle, and end of the school year in order to gauge the effectiveness of the new course *Mobile Technology*. Insightful student perspectives help to provide evidence of *Mobile Technology*'s ability to benefit student learning.

Student Perspectives on Mobile Learning

At the end of the school year, 48 middle school students enrolled in Motter's four sections of *Mobile Technology* reflected on their mobile learning experiences while completing an end of the year survey. All of the surveyed students were White and lived within the rural school district. Survey participants consisted of 21 female students and 27 male students. The survey reveals students' perspectives on mobile learning, their mobile technology usage outside of school, their comfort- and confidence-level with mobile devices and apps, their intentions to act as responsible digital citizens, and their plans to continue their use of mobile technologies in the future. As demonstrated by the following student responses, most surveyed students reported positive feedback regarding their mobile learning experiences in, and outside of, *Mobile Technology*.

Ninety-two percent of surveyed eighth graders enrolled in *Mobile Technology* indicated that using mobile devices enhanced their learning experience during the 2015–2016 school year (see Figure 10.1). In addition to finding mobile devices as beneficial learning tools, students seemed to favor using the mobile app Google Classroom in the classroom. Eighty-one percent of surveyed eighth graders enrolled in *Mobile Technology* said using Google Classroom, for the first time this school year, enhanced their learning experiences in school (see Figure 10.2). Additionally, 54% of students feel very comfortable and confident using Google Classroom as an educational tool (see Figure 10.3).

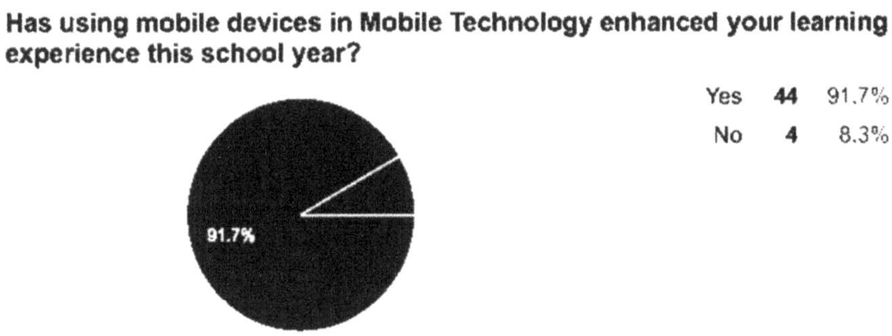

Figure 10.1. Student perspectives on *Mobile Technology*. This figure represents students' end of the school year perspectives on their learning experiences in *Mobile Technology*.

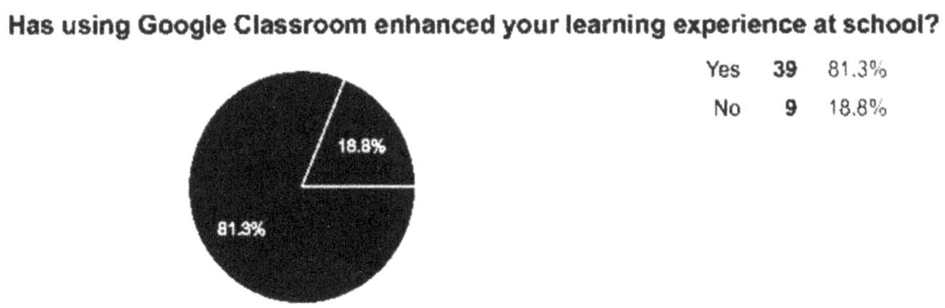

Figure 10.2. Student perspectives on Google Classroom. This figure represents students' end of the school year perspectives on using Google Classroom in school.

How comfortable and confident are you using Google Classroom?

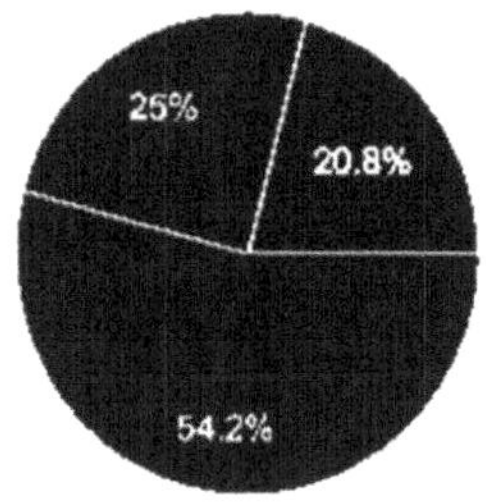

I am very comfortable and confident	**26**	54.2%
I am comfortable and confident	**12**	25%
I am somewhat comfortable and confident	**10**	20.8%
I am uncomfortable and unconfident	**0**	0%
I am very uncomfortable and unconfident	**0**	0%

Figure 10.3. Student confidence Using Google Classroom. This figure represents students' comfort- and confidence-level using Google Classroom at the end of the school year.

Middle school students' mobile learning experiences in school seem to impact their use of mobile technologies outside of school. Sixty-nine percent of surveyed eighth graders enrolled in *Mobile Technology* also reveal they use apps they learned with in *Mobile Technology* while at home (see Figure 10.4). Students seem to be more mindful of their actions and interactions online during and outside of class, as 63% of surveyed students seem to believe they are more cautious and better digital citizens (see Figure 10.5). Understanding how students use mobile technologies outside of school can assist teachers with mobile technology curricular decisions (Hughes et al., 2015). Finally, taking *Mobile Technology*, seemed to influence the majority (81%) of eighth graders to continue using mobile technologies in the future (see Figure 10.6).

Survey results reveal that the majority of surveyed *Mobile Technology* students believed they benefited from using mobile devices and mobile apps in the classroom and beyond. This overwhelmingly positive student feedback provides evidence that *Mobile Technology* was successfully in effectively enhancing middle school students' learning experiences in school while guiding their development of 21st century knowledge and skills.

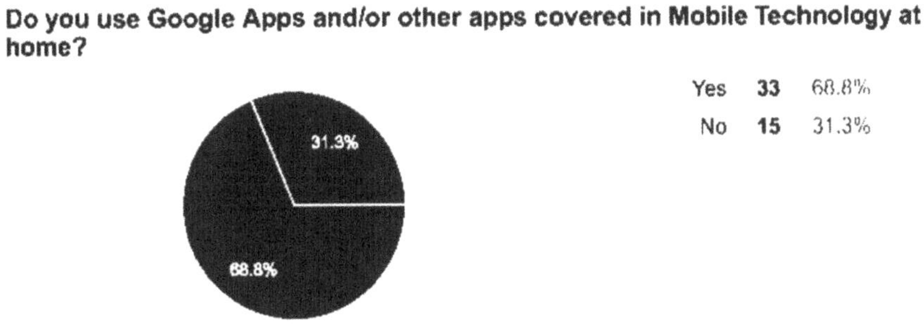

Figure 10.4. Student app usage at home. This figure represents student usage of apps covered in the classroom at home.

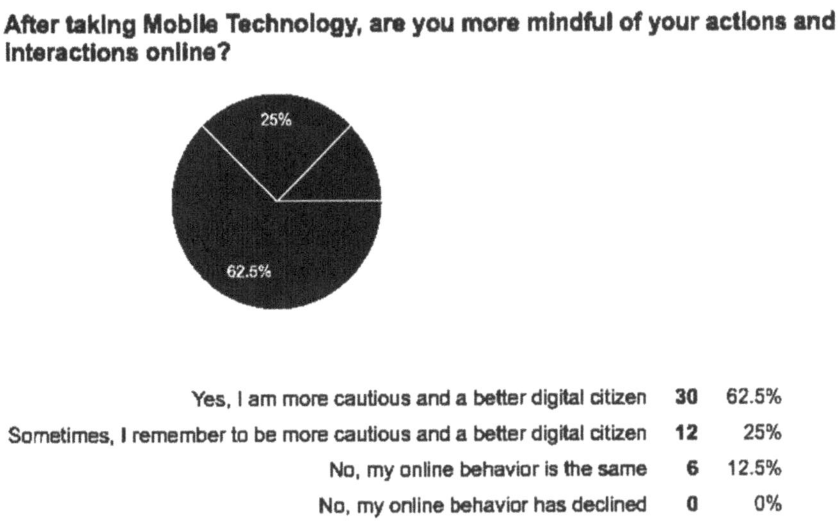

Figure 10.5. Student online behavior. This figure represents students' perspectives on their online behavior after participating in *Mobile Technology*.

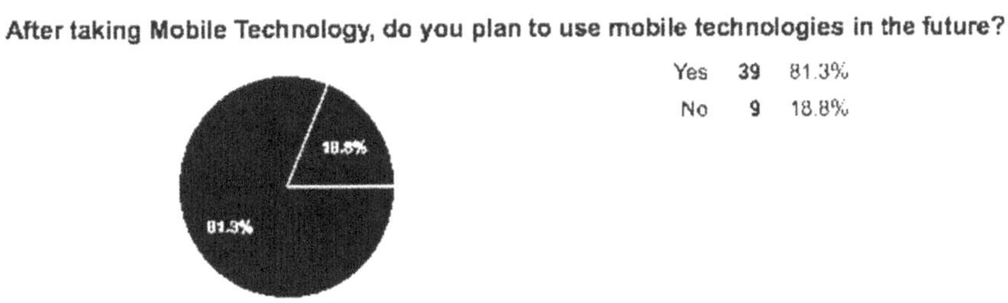

Figure 10.6. Future student use of mobile technologies. This figure represents students' intentions at the end of the school year to use mobile technologies in the future.

CHALLENGES OF USING MOBILE TECHNOLOGIES AS INSTRUCTIONAL TOOLS

The *Mobile Technology* course provides evidence that new media and mobile technologies can effectively facilitate teaching and learning, if those using them are motivated to teach and learn with contemporary instructional tools. However, Motter found that mobile technologies may also present unforeseen obstacles. Schools may struggle with digital technology integration if they serve a large number of at-risk students, English language learners, or students, such as those living in rural communities, with limited outside-of-school technology exposure. At Flood Valley, the time constraints of standardized testing also limited digital technology integration in tested subject areas (Hughes et al., 2015).

Access and Experience

In this next section, examples from the case are used to suggest challenges schools face when implementing mobile technologies as instructional tools. As Motter found, requesting new technology, such as mobile devices, can be frustrating for teachers at top-down schools where department chairs or administrators who must approve technology requests may be resistant to technology, or do not share a teacher's educational vision. As demonstrated by the *Mobile Technology* course, in order for teachers to effectively integrate mobile technologies into the curriculum, they need to be supported by others and provided the necessary tools. A technology planning committee could function as a support system for teachers by collectively developing an evaluation system for mobile technologies and considering the feedback of teachers, students, the public, and administrators to ensure it is successfully implemented (Picciano, 2011).

Additionally, educators should be aware of the challenges associated with new media and mobile technologies such as lack of equal access, experience, and training and programming skills, particularly for minority groups such as women (Bautista, 2003; Keifer-Boyd, 2007). While planning for and instructing a mobile learning course, Motter encourages others to be mindful of these issues and strive to provide equal opportunities for all students in physical and virtual learning environments while utilizing mobile technologies.

Time

Motter found that facilitating meaningful mobile learning during short class periods can also be challenging for instructors. In some instances, *Mobile Technology* instructors inaccurately anticipated the amount of time it would take to cover content in *Mobile Technology*. Due to limited time, students did not experience everything originally planned for the course. For example, Motter intended to use Lenovo Yoga tablets with students to explore Knowledge Matters, Virtual Business Retailing (https://vb.knowledgematters.com/), which involves an online self-guided unit interweaving reading, math, and simulations with business retailing curriculum. Educators need to be prepared, yet flexible when planning for learning with mobiles.

Anonymity

Similar to a physical classroom, virtual space can allow for negative actions. Motter believes it is important for instructors to discuss ethical behavior in virtual space with students (Julian, 1997) along with issues that may accompany identity (mis)representation that is possible in virtual environments. Anonymity in virtual space allows for slanderous actions that may negatively impact both the participant and public audience. While instructing *Mobile Technology*, Motter served as an online content moderator. This would serve to address criticisms of online environments based on privacy issues and seemingly unmonitored behavior (Overby, 2009).

Motter found that when engaging students in online environments, students' security and privacy must be considered. Concealing students' identity when posting online may be necessary to help reduce the potential for cyberbullying. The possibility for acts of hatred to take place in the virtual world and the physical world exist, but instructors who act as site moderators can aid in reducing harmful dialogue.

CALL FOR FUTURE DIGITAL LEADERSHIP AND PRACTICE IN PRE-K–12

This chapter concludes by calling for other educators interested in serving as digital leaders to meaningfully interweave mobile technologies into curricula in order to support learning goals and the advancement of students' 21st century knowledge and skills. By interweaving, rather than integrating, mobile technology we are seamlessly utilizing it in daily classroom routines, enabling it to become a natural tool and naturalized part of the learning process in educational environments (Saltman, 2014). Effective interweaving of mobile technologies cannot occur without a coherent vision for implementation and the support of all stakeholders including teachers, parents, students, school board members, and administrators (Sheninger, 2014). A clear plan must be developed that addresses why and how mobile technology will be used as classroom learning support. When purchasing mobile technologies, reliability, cost, accessibility, capacity, and user-friendliness should be considered. School districts and administrators can support meaningful mobile technology implementation in pre-K–12 by first conducting school site research, and then providing teachers with adequate resources, time, support, input, and continuous up-to-date training. Educators should engage in self-reflexive practice when determining if mobile technologies are being meaningfully and smoothly interwoven throughout the school year. Additionally, mobile technology implementation needs to be evaluated by teachers and administrators to ensure its effectiveness in enhancing student learning (Picciano, 2011).

Digital leaders possess the power to make small changes over time that will have a lasting impact on school culture. "Transforming schools into digital-age places of learning requires leadership from people who can accept new challenges and embrace new opportunities, which is at the heart of digital leadership" (Sheninger, 2014, p. 71). Successful digital school leaders can motivate and empower students and colleagues via effective implementation of instructional mobile technologies that promote empowered self-directed life-long learners (Svinicki & McKeachie, 2011; Walvoord & Anderson, 2010).

By conveniently providing access to vast and rich resources, mobile devices change how and when people learn. Mobile technologies have a uniquely powerful potential to captivate interest and motivate learners' engagement across the curriculum. Providing proper guidance of mobile technology usage can positively transform student learning experiences. We cannot expect students to act as responsible digital citizens outside of school if we fail to effectively interweave meaningful mobile learning experiences in the classroom.

REFERENCES

Apple Inc. (2016). Keynote (3.0.5) [Mobile application software]. Retrieved from http://itunes.apple.com

Autodesk Inc. (2016). Tinkerplay (1.2.3) [Mobile application software]. Retrieved from http://itunes.apple.com

Bautista, R. O. (2003). Staking their claim: Women, electronic networking, and training in Asia. In G. Dines & J. M. Humez (Eds.), *Gender, race, and class in media: A text-reader* (2nd ed., pp. 708-714). Thousand Oaks, CA: SAGE.

Brainshark Inc. (2016). SlideShark (4.3.1) [Mobile application software]. Retrieved from http://itunes.apple.com

Butrymowicz, S. (2014). Bridging the digital divide in America's rural schools. *The Hechinger Report*. New York, NY: Teachers College at Columbia University.

Retrieved from http://hechingerreport.org/bridging-the-digital-divide-in-americas-rural-schools/

Cateater, LLC. (2016). Stop Motion Studio (8.0) [Mobile application software]. Retrieved from http://itunes.apple.com

Cegnar, J. (2016). OSnap! (3.4) [Mobile application software]. Retrieved from http://itunes.apple.com

Common Sense Media. (2016a). *Copyright and fair use animation* [Video file]. Retrieved from https://www.commonsensemedia.org/educators/lesson/copyrights-and-wrongs

Common Sense Media. (2016b). *Scope and sequence: Common sense K–12 digital citizenship curriculum* [Unit plans]. Retrieved from https://www.commonsensemedia.org/educators/scope-and-sequence

DNews. (2014, February 16). *Emoticons are changing our brains!* [Video file]. Retrieved from https://www.youtube.com/watch?v=5htUgnSfg70

Dolan, J. E. (2016). Splicing the divide: A review of research on the evolving digital divide among K–12 students. *Journal of Research on Technology in Education, 48*(1), 16–37.

Goodwill Community Foundation, Inc. (2016). Tutorial: Windows 10 [Tutorial]. Retrieved from http://www.gcflearnfree.org/windows10/

Google, Inc. (2016). G Suite for Education (2.2016.48201) [Mobile application software]. Retrieved from https://www.google.com/edu/products/productivity-tools/

Hall, B.S. (2013). Smartphones have bridged the digital divide. *Readwrite*. Retrieved from http://readwrite.com/2013/05/17/smartphones-have-bridged-the-digital-divide/

Harper, B., & Milman, N. B. (2016). One-to-one technology in K–12 classrooms: A review of the literature from 2004 through 2014. *Journal of Research on Technology in Education, 48*(2), 129–142.

Hern, A. (2015). Don't know the difference between emoji and emoticon? Let me explain. *The Guardian*. Retrieved from https://www.theguardian.com/technology/2015/feb/06/difference-between-emoji-and-emoticons-explained

Herold, B. (2015, November). Stuck with huge bills for Internet service that barely works Calhoun County schools have missed out on the digital revolution. *Education Week*. Retrieved from http://www.edweek.org/ew/projects/2015/rural-schools-broadband/the-slowest-internet-in-mississippi.html

Hokanson, K. (2015). Conquering copyright confusion is fundamental to digital citizenship. *Entrsekt, 2*(1), 10–11.

Hughes, J. E., Read, M. F., Jones, S., & Mahometa, M. (2015). Predicting middle school students' use of Web 2.0 technologies out of school using home and school technological variables. *Journal of Research on Technology in Education, 47*(4), 211–228.

International Society for Technology in Education. (2007). ISTE standards for students [PDF]. Retrieved from https://www.iste.org/docs/pdfs/20-14_ISTE_Standards-S_PDF.pdf

Ito, M., Horst, H., Bittanti, M., Boyd, D., Herr-Stephenson, B., Lange, P. G., Pascoe, C. J., & Robinson, L. (2008). *Living and learning with new media: Summary of findings from the digital youth project.* Chicago, IL: MacArthur Foundation.

Retrieved from http://www.macfound.org/atf/cf/%7BB0386CE3-8B29-4162-8098-E466FB856794%7D/DML_ETHNOG_WHITEPAPER.PDF

Jenkins, H., Purushotma, R., Clinton, K., Weigel, M., & Robinson, A. J. (2007). *Confronting the challenges of participatory culture: Media education for the 21st century.* Chicago, IL: The MacArthur Foundation. Retrieved from http://digitallearning.macfound.org/atf/cf/%7B7E45C7E0-A3E0-4B89-AC9C- E807E1B0AE4E%7D/JENKINS_WHITE_PAPER.PDF

Julian, J. (1997). In a postmodern backpack: Basics for the art teacher on-line. *Art Education, 50*(3), 23–42.

Keifer-Boyd, K. (2007). Cyberfeminist activist art pedagogy (Park, Jeong Ae, Trans.). *Journal of Research in Art & Education, 8*(1), 29–89.

Keifer-Boyd, K., Amburgy, P. M., & Knight, W. B. (2007). Unpacking privilege: Memory, culture, gender, race, and power in visual culture. *Art Education, 60*(3), 19–24.

Kim, P., Hagashi, T., Carillo, L., Gonzales, I., Makany, T., Lee, B., & Gárate, A. (2011). Socioeconomic strata, mobile technology, and education: A comparative analysis. *Education Technology Research and Development, 59*(4), 465–486.

Léger, M. T., & Freiman, V. (2016) A narrative approach to understanding the development and retention of digital skills over time in former middle school students, a decade after having used one-to-one laptop computers. *Journal of Research on Technology in Education, 48*(1), 57–66.

Liao, C. L. (2008). Avatars, Second Life, and new media art: The challenge for contemporary art education. *Art Education, 61*(2), 87–91.

Lister, M., Dovey, J., Giddings, S., Grant, & Kelly, K. (2009). *New media: A critical introduction* (2nd ed.). New York, NY: Routledge.

McMunn Dooley, C., Lewis Ellison, T., Welch, M. M., Allen, M., & Bauer, D. (2016). Digital participatory pedagogy: Digital participation as a method for technology integration in curriculum. *Journal of Digital Learning in Teacher Education, 32*(2), 52–62.

Nair, S. (2016). Transforming digital users into digital creators. *Entrsekt, 2*(4), 16

Overby, A. (2009). The new conversation: Using weblogs for reflective practice in the studio art classroom. *Art Education, 62*(4), 19–24.

Partnership for 21st Century Skills. (2007). Framework for 21st century learning. Retrieved from http://www.p21.org/our-work/p21-framework

Pew Research Center. (2017a). Internet/broadband fact sheet [Fact sheet]. Retrieved from http://www.pewinternet.org/fact-sheet/internet-broadband/

Pew Research Center. (2017b). Social media fact sheet [Fact sheet]. Retrieved from http://www.pewinternet.org/fact-sheet/social-media/

Picciano, A. G. (2011). *Educational leadership and planning for technology* (5th ed.). Upper Saddle River, NJ: Pearson Education.

Prezi, Inc. (2017). Prezi Viewer (4.3.2) [Mobile application software]. Retrieved from http://itunes.apple.com

Saltman, D. (2014). Jennifer Motter: Interweaving technology and art. In D. Saltman (author), *Teachers talking tech: Creating exceptional classrooms with technology* (pp. 45–59). Cambridge, MA: Harvard Education Press.

Scheffer, J. L. (2016). Leverage technology to give students a voice, choice in their learning. *Entrsekt, 2*(4), 8–9.

Sheninger, E. (2014). *Digital leadership: Changing paradigms for changing times*. Thousand Oaks, CA: Corwin.

Smith, C. A., & Santori, D. (2015). An Exploration of iPad-based teaching and learning: How middle-grades teachers and students are realizing the potential. *Journal of Research on Technology in Education, 47*(3), 173–185.

Svinicki, M., & McKeachie, W. J. (2011). *McKeachie's teaching tips: Strategies, research, and theory for college and university teachers* (13th ed.). Belmont, CA: Wadsworth, Cengage Learning.

Thielman, S. (2016). How a digital divide leaves parts of rural America isolated. *The Guardian*. Retrieved from https://www.theguardian.com/technology/2016/may/16/rural-america-internet-access-navajo-nation

Walvoord, B. E., & Anderson, V. J. (2010). *Effective grading* (2nd ed.). San Francisco, CA: Jossey-Bass.

Wheeler, T. (2014). Closing the digital divide in rural America. Washington, DC: Federal Communications Commission. Retrieved from https://www.fcc.gov/news- events/blog/2014/11/20/closing-digital-divide-rural-america

White House. (2015). Mapping the digital divide. *Council of Economic Advisers Issue Brief*. Retrieved from https://www.whitehouse.gov/sites/default/files/wh_digital_divide_issue_brief.pdf

CHAPTER 11

BYOD IN THE ART CLASSROOM

A Framework for Studio Learning

Aysenur Ozyer
University of Colorado Denver

Kyle Roberts
Cherry Creek School District, Castle Rock, CO

Brent G. Wilson
University of Colorado Denver

Art classrooms in high schools and colleges are places of hands-on practice and construction, very much in keeping with today's emphasis on active, experiential learning. With support from occasional guidance from the instructor, students use class time to create works of art in accordance with a standards-based curriculum plan. Students also review and critique each other's work and offer help to peers nearby. Art classrooms bear a close resemblance to other production-oriented learning environments such as robotics labs and digital design studios. These learning settings focus on creative expression, student autonomy, master-apprentice mod-

Mobile Learning: Perspectives on Practice and Policy
pp. 231–246

eling, and learn-by-doing activities. As an outpost of creative, constructive learning, art classrooms are often a refuge for students at risk of quitting school or otherwise disengaging from the system.

In this chapter, we examine the art classroom as a critical case for studiolike pedagogies with students accessing resources via mobile tools. We show how this approach suits art classroom especially well and plays a distinctive role in the larger curriculum. Beginning with an overview of smart devices and their uses for learning, we offer a simple pedagogical framework, illustrated by a case description of coauthor Kyle Roberts' classroom. Although focused on art curriculum, many of the strategies carry promise for other elective subjects at the high-school level.

MOBILE LEARNING AND BRING-YOUR-OWN-DEVICE FLEXIBILITY

Most teenagers have access to a smartphone (Anderson, 2015). Indeed, smartphones can help underserved students connect to information and web resources. Lenhart, Ling, Campbell, and Purcell (2010) report that 21% of teens have only their cell phones to access the Internet. More teens from lower income homes are accessing the Internet primarily via their phones: 44% of Black teens and 35% of Hispanics use their phones for primary online access, compared to 21% of White teens (Lenhart et al., 2010).

In spite of occasionally prohibitive school policies, use of mobile devices in classrooms is increasing (Herro, Kiger, & Owens, 2013; Leinonen, Keune, Veermans, & Toikkanen, 2014; Ng, 2015). This is encouraging, because mobile devices hold enormous potential for learning and instruction. Sharples (2000) lists several affordances of mobile devices for learning including portability, personalization, availability, accessibility, adaptability, persistence, and convenience. Cochrane and Bateman (2010, p. 2) also add innovation and equity to the list. Consistent with emerging research, the classroom described in this chapter utilizes mobile devices to:

- improve accessibility (Squire & Dikkers, 2012);
- empower students by giving them the opportunity to research and choose their own personalized content (Squire & Dikkers, 2012);
- increase motivation and interest by having students internalize what they do (Herro et al., 2013; Norris & Soloway, 2008; Squire & Dikkers, 2012; Swan, Hooft, Kratcoski, & Unger, 2005);
- foster learning of 21st century skills like collaboration, inquiry, and communication (Herro et al., 2013; Leinonen et al., 2014; Norris & Soloway, 2008);

- improve students' academic achievement by learning to use the Internet for research (Kiger, Herro, & Prunty, 2012); and
- increase quality and quantity of student work and extend learning beyond the classroom (Swan et al., 2005).

Although the research is encouraging, financially strapped schools struggle to provide their students with web-connected devices, mobile or otherwise (Ng, 2015). Bring-your-own-device (BYOD) is defined as "technology-mediated learning where students bring a personally owned smart mobile device from home to school for the purpose of learning" (Ng, 2015, p. 173). With the help cost- and resource-sharing policies and classroom practices, schools can benefit from mobile learning while connecting to outside resources more naturally.

ACTIVE LEARNING

Herro et al. (2013) posit that as mobile device usage grows, the users' demands of the hyperconnected, affordable, and global access will disrupt schools and cause a paradigm shift where education is perceived as "I learn" rather than "we teach." This more student-centered view will pave the way for *active learning*, which is defined as an instructional method where students engage and participate in their learning with meaningful learning activities while thinking about what they are doing (Prince, 2004). Leu et al. (2009) report "although the digital generation may be skilled with the new technologies, children often lack critical thinking skills for evaluating information obtained online" (as cited in Gormely & McDermott, 2011, p. 14). This parallels Lei's (2009) finding of digital natives often lacking that learning or educational dimension. What coauthor Kyle Roberts tries to do in his student-centered arts classroom is consistent with active learning principles. Students are encouraged using their connected mobile devices to research, learn, and find solutions, in the process internalizing and personalizing their learning. Mr. Roberts motivates his students to actively seek solutions and evaluate information they find online hence promoting critical thinking and problem solving (Ackermann & Krupp, 2012; Prince, 2004; Thomas, Singh, Klopfenstein, & Henry, 2013; Tune, Sturek, & Basile, 2013). Guiding students through defining questions, searching for supplementary information using mobile devices, solving problems via collecting evidence on the Internet, summarizing findings and making conclusions–these activities will help students develop their 21st century skills (Berk, 2009; Gaughan, 2014; Hwang, Lai, & Wang, 2015, p. 459).

FLIPPED LEARNING

When students are active learners taking more responsibility for their learning, higher levels of educational objectives can be achieved (Hwang et al., 2015). Bloom's taxonomy (Anderson & Krathwohl, 2001) defines these higher levels as applying, analyzing, evaluating, and creating. However, when these higher levels of objectives are sought, the time needed for instruction often increases. The flipped classroom can be defined as "a pedagogical approach which moves the learning contents taught by teachers' direct instruction to the time before class in order to increase the chances for the students and teacher to interact" (Hwang et al., 2015, p. 452). Traditional classroom environments tend to favor whole-class lectures targeting lower levels of Bloom's taxonomy, remembering and understanding, while students are left alone to achieve the top levels, analyzing, evaluating, and creating (Bormann, 2014). On the other hand, in the case of flipped classrooms, more remember-level content is delivered outside of the classroom, enabling students and teaching more time to focus on higher levels in class (Bormann, 2014). "Students would focus on and engage in interactive activities that foster a deeper understanding of the content through discussion, practical work, and problem-solving tasks that they do individually or collaboratively in small groups" (Ng, 2015, p. 150).

In the case detailed below, Mr. Roberts provides students with guiding materials before the classroom so they can exercise the lower level of objectives, remembering and understanding, at home. Therefore, students have more time in the classroom to analyze, evaluate, and create thus fostering self-directed learning combined with reflection and taking responsibility for learning (Bormann, 2014; Hwang et al., 2015; Strayer, 2012). Meanwhile, Mr. Roberts acts as the facilitator who guides learning, assesses student needs, and provides individual assistance in class (Ng, 2015). In this way, Mr. Roberts's classroom can focus on: creating or providing high-quality instructional materials and content for students to use; developing high order thinking tasks; and designing interactive and engaging course environments where students can apply what they learned to real world problems for the classroom time (Hwang et al., 2015).

ART STUDIO: A FRAMEWORK

The research and literature cited above provide a useful basis for approaching classroom pedagogy in the art classroom, but they need synthesis into a coherent approach. The key elements of our proposed synthesis are outlined below:

- *Overall Goal: Creative Production.* To create a meaningful visual work of art, well received by a defined audience and applying sound standards-based skills and principles (National Coalition for Core Arts Standards, 2014).
- *Means: flipped pedagogy.* Adopt a project-driven pedagogy that provides an avenue for greater student control and autonomy through inquiry and interest-driven pursuit, requiring full student attention, engagement, and investment.
- *Environment: learning studio.* Provide a comfortable, task-oriented but casual and aesthetically pleasing environment with enough space for creative and productive work.
- *People: Secure but Challenging Relationships.* Foster a social climate where learners benefit from peer and instructor support and as well as constructive critique and guidance.
- *Accessible Tools and Resources.* Through BYOD, ensure access to a rich variety of resources, including teacher-created and generic videos, demos, and visual images–available prior, during, and after class.

Each element in the framework, we believe, is critical—if any are missing, productivity breaks down or is disrupted. In that sense, they are systemically connected, with the overall approach being more than the sum of its parts.

MR. ROBERTS' ART CLASSROOM

This section presents the art classroom of Mr. Roberts (coauthor Kyle) as a concrete case illustrating the framework. The narrative draws on a 1-day class visit by the research team and an earlier draft report (Roberts, 2015). In February 2016, the researchers visited Mr. Roberts's high school, a 15-year-old school with more than 2,500 students. This visit aimed to observe Mr. Roberts's class. Classroom observation helped researchers to better understand the dynamics of the instruction and student engagement. Mr. Roberts's class was a second-year elective with a range of beginning and continuing art students. The class was primarily White reflecting the middle-class neighborhood and suburban district.

Mr. Roberts has been teaching high-school art for 7 years, three at the present school. He reflected on his students and instruction:

> Over my career, I have seen a problem that persists across mediums and levels: Students of higher need will tend to dominate my time, leaving my students with drive and talent underappreciated and eventually uninterested.

> Students in every class exhibit a range of background knowledge and skill; some attend with the mindset of fulfilling a requirement while others are truly pursuing their passions. The net effect of these diverse needs and purposes, coupled with the limitations of a single teacher, create a clear need for more flexible access to expertise and guidance.
>
> I came to the conclusion that the best way to strengthen everyone's performance and experience in the class was to develop a more flexible environment along with a BYOD policy in my classroom. In my school, currently I have been supported from my administration in my endeavors to increase student performance with the use of devices in the classroom. This practice varies from subject to subject; however, it seems especially prevalent in elective classes like mine.

Mr. Roberts is aware of challenges posed by high-stakes testing and limited funding, yet believes it is very important to draw students into the arts program. His goal is "for every student to not only be challenged but also create meaningful/successful products and be inspired to take more art classes" (Roberts, 2015, p. 1). Limited class time poses another challenge since it is not always possible to personally work with individual students, answer their questions, and guide them through their projects. Mr. Roberts continues:

> I developed some online instructional material that was accessible to students whenever and wherever they needed it. Students are encouraged to bring and use their device in class to help with their development. A cart of laptops is also available for those who need them. Especially in areas where I lacked expertise, students could use YouTube or other platforms to find the help they need. Students are thus acquiring skills for helping themselves, which they can continue at home or other settings and apply to different problems and needs in their schooling.

The Learning Studio

Mr. Roberts' classroom simulates an art studio, with learning focused on individual student projects supported by demos and minilectures. He adopts BYOD policy to facilitate student-initiated access to resources. BYOD support fills important needs, specifically (1) assuring 1:1 access with provided laptops complementing student-owned phones and laptops, and (2) providing students with comfortable, seamless, and familiar means of accessing resources before, during, and after class. Students give each other limited support at a table, with additional support by Mr. Roberts as he circulates among tables. This studio-style classroom allows him to focus extra energy toward specific students with special needs.

Table 11.1. Student Learning Activities Linked to Goals and Tools

Activity	*Goal*	*Tools/Resources*
Web searches • Images • Videos • Artists • Subject matter	• Ideas and inspiration • Reference pictures • Media elements (for inclusion in projects) • Information access	• Search tool (Google) • Image banks
Demos and mini-lectures • Instructor (in-class and online) • Generic (e.g. YouTube, Khan Academy)	• Guidance and skill-building	• YouTube • Tutorial sites (e.g., Kahn Academy)
Digital image manipulation	• Create a reference picture • Designing/problem solving • For future work	• Adobe Photoshop • Other tools and apps
Collaboration/communication • Peers • Instructors • Outsiders	• Informal/formative critiques • Final critique	• LMS • Communication tools • Social media
Reports • Written notes/plans	• Analysis and reflection	• Word processors • Document/ PDF apps • Printers
Sharing/portfolios • Within school • Outside	• Learning through creative conversation/expression • Recognition and support through wards/competitions	• Website development tools

To make in-class time more efficient and personalized, Mr. Roberts uses flipped instruction. He uploads prerecorded step-by-step how-to videos and documents to the school's learning management system and YouTube. He instructs students to watch the videos and review the support documents before class, and encourages them to practice at home and rewatch any areas they did not grasp the first time. He utilizes apps to send text reminders to students and demonstrates how to access the content in the classroom. Use of smartphones and other mobile devices in the classroom is allowed and encouraged under certain conditions. Rules for technology use in the classroom are explained to students (Roberts, 2015). Table 11.1 presents a summary of learning activities in Mr. Rob-

Figure 11.1. Mr. Roberts's sketch depicting his art classroom and student learning.

erts's classroom, with links to learning goals and specific tools used in the activity. A visual depiction of Mr. Roberts's classroom can also be seen in Figure 11.1.

During class time, students routinely access the Web for educational purposes. Students have opportunities for collaboration at the table, but each student is required to complete an individual creation of art. This brings a level of accountability that some collaborative environments lack–namely that each student will personally demonstrate core skills and competencies. Students at tables exercise considerable agency and responsibility for their individual work. Within project parameters, they choose subject matter, develop plans and ideas, search the Web for guidance and video demos, and seek help and advice from peers and instructor. Because expertise in the arts relies on aesthetic judgment, students develop both technical skills and a sense of style and judgment for using those skills. This happens with experience gained through repeated skills practice at increasing levels of sophistication.

Everyday Activities

We report in this section more detail about routine learning activities in the learning studio. In Mr. Roberts's description:

> Combining BYOD and the laptop cart, everyone has comfortable access to technology. Students who have questions, who missed school, who work more quickly than others or who just need to see the demo again, are using their device to help with their learning experience. From the instructor side, it is quite useful to have this online resource acting almost as another instructor at times.
>
> When we begin a new assignment in my class, students are presented with a new skill and creative problem for their assignment. Students are presented with how they can solve the problem and what check points I will be looking for in the grading rubric. Usually the prompts presented are wide ranging. They allow students of all skill levels to achieve their goal while pushing their individual talents to new levels. After my prompt has been presented, the next step usually will be students using their phones/computers to find reference pictures and generate ideas about how to solve the problem. Students will often go through the Internet mainly using Google and compile a bank of images that they will either reference or alter during their brainstorming session.

The classroom visit of the researchers confirmed this report. The overall "vibe" was relaxed and focused: Students enjoyed considerable autonomy as they worked at their tables. Some consulted together and helped each other with their projects. Others worked alone but seemed open to input from Mr. Roberts or others. One student kept his head on the table–we were told he needed rest, and was not productive that day. When we engaged individuals or small teams, students were positive and articulate about their goals and activities, and open to feedback and suggestions. Mr. Roberts spent about 10% of classroom presenting to the full group–the remainder consulting with individuals and tables.

Mr. Roberts continues his description of studio activity:

> Another thing that is recommended for students is to take advantage of the classroom camera and our department's photo lab, where students will often use their peers as models. Once these pictures are taken they are often put into Photoshop for more altering and then they are printed off and used for their paintings. The digital steps are also accompanied by students using paper and pencil to sketch out ideas for their work.
>
> Before a student can begin their final draft of a work they need to run it by me for approval. This is where I can catch students who may be on the wrong track or just need a little altering. Artists encountering problems will often talk to their peers or they might use Photoshop to see possible ideas. Or a student might reference a tutorial on YouTube that addresses how to

achieve a certain style or technique. When a student finishes their project, they will then turn it in to me, along with a self-reflection, for final critique.

Mr. Roberts continued by providing an example of a student work from his classroom:

One student, whom I will call Kate, was looking on her phone at Instagram and found a painter whose style she liked and wished to replicate. After studying and narrowing down this style, Kate then took pictures of her friend to use as a model and uploaded them to Photoshop. The photos were then altered in a manner that resembled the style in question. Next Kate projected the picture onto a canvas and traced the outline. Once this out line was there, she painted the picture over the next couple of weeks. When the figure was painted, Kate ran into questions about what to do with the background. She and her friend then went back to Photoshop for ideas and ultimately came to a solution. Armed with this, they painted the background and finished the painting. This painting was recently submitted to a state competition and accepted for showing. The competition receives over 6,400 entries and chooses less than 10% of total works. With the acceptance of the artists' work they will now be able to apply for a scholarship to help them with their schooling.

Learning and Teaching Benefits

The learning studio accrues important benefits for both teacher and students:

I have observed greater equity in how students are treated. There will always be some students requiring more attention from me, leaving others at higher risk of quiet neglect. Now with access to Web resources, everyone feels more capable of finding ideas and support themselves. That lessens the guilt I feel sometimes, when I'm not sure I've allocated my time and attention in the right way. Achieving learning isn't just on my shoulders, it's a shared responsibility.

I also notice an impact on the risks I take pedagogically. I now offer more open-ended assignments, sometimes with a focus on learning a demanding technical skill, other times when I look for new ideas and synthesis of skills. I can make adjustments along the way, uploading tutorials or referencing other students' work as examples. This attitude of sharing moves me away from top-down control and toward a role of ground-level facilitator. Students have more direct, first-hand experience with a structured creative process–the kind used by successful professional artists.

With access to technology, my students have become more empowered. Mobile devices have opened up the sharing of ideas and information that have allowed more learners to thrive. Once students get used to the open,

> sharing environment, I see more engagement and time on task, and less exclusive reliance on the guidance I provide. When final products are created and presented, I tend to see greater variation and creativity.

Please refer Figure 11.1 for a sketch depicting Mr. Roberts's art classroom and student learning.

Other Classes

Mr. Roberts describes below how the studio approach works in other classes, more advanced and beginning level.

> I have another set of classes where students are working at the top levels of my instruction. This class is comprised of the brightest minds within the school and many of them are the top artists as well. This is an extremely demanding class to teach due to the individual instruction necessary for the students to flourish. Because of this, the instructor needs to be aware of and competent in a multitude of media and knowledgeable of art history/current work. Students in this class set out to explore their own ideas and artistic styles that inspire them. This is where the use of technology has become so important for these students. Without access to these devices, students would be limited to a small amount of resources that would impact the reach and quality of their art. This technology use is prevalent during the creation of every type of art within the class and is not limited only to only one medium.
>
> In my entry-level classes I am faced with teaching students of varying levels of skill and interest. Their diversity of needs presents a challenge to how I use my time in class. With the implementation of a learning studio I can reach more students with one-on-one guidance. These flipped lessons allow students to access tutorials and resources at any time without pulling me away or waiting for my attention. The YouTube counter on one of my videos is up to 75 views since becoming active earlier this year. This tutorial is 11 minutes long. Doing the math I calculate 800 minutes of instruction saved (or gained, from the student's view) within two semesters of use. Those minutes are a kind of metaphor pointing to improved teaching efficiency and improved performance of my students.

As noted in Kyle's detailed description of his classroom, he strives to create an atmosphere and learning that resemble a learning studio. Next, we apply our five framework elements. Mr. Roberts achieves the hoped-for learning studio as outlined below.

- *Overall Goal: Creative Production.* Assignments are projects aimed at creating products to be shared with defined audiences (in-class at first and then extending to a wider reach). Efficiency and technique are emphasized as much as free and creative expression–indeed the creativity depends on and grows out of the constraints and forms required in a given assignment.

- *Means: Flipped Pedagogy.* Mr. Roberts' classroom is guided by a progressive, project-based pedagogy, with most class time devoted to individual student performance. "Student-centered" may not be the right description, however, since assignment parameters are instructor-determined in accordance with state and national standards and the demands of the discipline and the work is done primarily in a collaborative setting.
- *Environment: Learning Studio.* The physical environment reflects both the cleanliness and order expected in a school classroom but also a comfortable, casual feel denoting a secure, home-like feel—a place where students want to be. The configuration of tables, artwork, and physical tools and resources is conducive to individual work but also allows for peer discussions and occasional instructor consulting. Whole-group demos and discussion are also available through projected media.
- *People: Secure but Challenging Relationships.* Through a combination of pedagogical method, instructor skills, and student buy-in, the class maintains a low-key focus on the work, with minimal disruption or off-task behavior. Through cycles of class sharing followed by constructive critique, students generally feel safe to present their work with the hope of improvement through feedback and review. The generally positive affective climate is an important outcome and contributor to the success of the class.
- *Accessible Tools and Resources.* Tools and resources play a support role in the work. That is, students could study and research a topic all day, and never get to the work of creating their own product. Given that need for balance, technology's role must be secondary. Mr. Roberts and students would agree that access to technology is critical, but only in that support role–not as a "star" unto itself.

As detailed above, the role of pedagogy and tools is interdependent: When online learning resources and BYOD policies are added to the project-oriented pedagogy and the studio environment, the level of student autonomy and responsibility grows. Students encounter flipped learning as a whole experience, with all the elements working together. As an elective, art class lets this happens without the formality of lengthy expository lectures or textbooks that might be needed in certain content- or exam-driven courses. Ironically, art students may be preparing themselves more directly for work outside of school when compared to more content-driven courses (e.g. core or AP classes): the goal is not a high score on a mandated exam but rather skill acquisition in a fairly authentic performance context using tools appropriate for the work.

IMPLICATIONS

Mr. Roberts' classroom adopts a flipped learning model with the help of mobile technologies where students take responsibility for their learning and exercise the 21st century skills. The personalized learning environment helps students internalize their learning. Higher level of learning objectives is achieved in the classroom since the time allows students analyze, evaluate, and create. Students are motivated since they own their learning and they are willing to attend the classroom. Using mobile technologies, they can access the material not only in the classroom but during their commute and in their homes. Providing a seamless, continuous learning experience (Hwang et al., 2015), mobile technologies combined with a flipped learning approach can extend learning for students from all subjects. Mr. Robert's approach relies on cooperation between students and the teacher using the flexibility provided by today's technologies. This shared responsibility helps students personalize and own their learning. Students engage, research, share, interact, participate, and create collaboratively in this accessible learning environment.

While art programs in general face threats of elimination to provide more time for high-stakes test preparation (Graham, Parker, Wilkins, Fraser, Westfall, & Tembo, 2002), the learning studio may provide a way to increase the perceived value in the eyes of both students and districts. Mr. Roberts's classroom utilizes mobile learning to engage students motivating them to enhance their creativity along with other skills. Moreover, the learning studio is consistent with the maker movement's focus on designing and making things as a means of learning complex subjects and skills. This active pedagogy stands in contrast to teacher-centric methods common in many core subjects in high school and college–less reliance on text and more on hands-on guidance and demonstrations.

We would hope that uses of the framework would reflect its origins in the art classroom, extending to similar elective classes. Students are generally there by choice in high schools, reflecting some intrinsic interest or at least a diversion from traditional academic subjects. The framework could be suitable for subjects not strictly bound or defined by mandated exams, texts, or narrow performance standards. Mr. Roberts's art class is an example of use of mobile technologies to spark creativity and interest in the subject while sharing the learning responsibility with students who are ready to step up and join as partners in their own learning. Indeed, hand-on and creative skills are best learned through a shared responsibility between instructor and student, with considerable flexibility given to defining project details. It is our belief that a parallel approach utilized in other elective subjects could maximize students' engagement and creativity.

CONCLUDING THOUGHTS

We live in "networked times" (Sweeny, 2008), surrounded by communication technologies in nearly every facet of our lives. It is only natural for the Internet to penetrate our classrooms. A carefully designed art curriculum that incorporates Internet and mobile devices seems natural and fitting. Sweeny calls on art instructors to take advantage of the Internet's flexibility as a research tool, interactive social space, communication channel, and art gallery. In a networked art classroom where students are encouraged to research, participate, and interact, new forms of creation come to surface. We agree with Sweeney that using the Internet in a meaningful way can improve artistic expression and creation, increase work-related interaction, and allow self-organization by offering students an abundance of directly accessible resources. Routine integration of technology in art classrooms lessens the isolation and furthers the crossover effect from classroom to "real life"—and vice versa. Art classrooms can be places where the roles of teachers and students dynamically and continually reshape their roles, products, and learning outcomes. Technology supported art classrooms will support future inquiry and help students learn art collectively (Castro, 2012).

To better prepare for the future, students need sustained practice using 21st century skills such as creativity, collaboration, and inquiry. A studio-based art classroom offers that kind of sustained practice with the help of mobile devices and online resources–a closer match to our plugged-in life outside the classroom. Instructional methods like those in our framework can help 21st century skills to take hold and flourish, especially if they reinforced in different courses and throughout the schooling experience. In a way, teaching these higher order skills is easier to put into practice in art classrooms than in traditional classrooms. We call on art instructors to capitalize on this flexibility and adopt practices in their classroom that help their students to prepare for what the future brings.

REFERENCES

Ackerman, A. S., & Krupp, M. L. (2012, October). *Five components to consider for BYOT/BYOD.* Paper presented at International Conference on Cognition and Exploratory Learning in Digital Age, Madrid, Spain.

Anderson M. (2015). *Technology device ownership: 2015.* Retrieved from http://www.pewInternet.org/data-trend/mobile/deviceownership/

Anderson, L. W. (Ed.), Krathwohl, D. R. (Ed.), Airasian, P. W., Cruikshank, K. A., Mayer, R. E., Pintrich, P. R., ... Wittrock, M. C. (2001). *A taxonomy for learning, teaching, and assessing: A revision of Bloom's Taxonomy of Educational Objectives* (Complete edition). New York, NY: Longman.

Berk, R. A. (2009). Multimedia teaching with video clips: TV, movies, YouTube, and mtvU in the college classroom. *International Journal of Technology in Teaching and Learning, 5*(1), 1–21.

Bormann, J. (2014). *Affordances of flipped learning and its effects on student engagement and achievement* (Unpublished master's thesis). University of Northern Iowa, Cedar Falls, Iowa. Retrieved from: http://fln.schoolwires.net/cms/lib07/VA01923112/Centricity/Domain/41/bormann_lit_review.pdfhttp:h

Castro, J. C. (2012). Learning and teaching art through social media. *Studies in Art Education, 53*(2), 152–169.

Cochrane, T., & Bateman, R. (2010). Smartphones give you wings: Pedagogical affordances of mobile Web 2.0. *Australasian Journal of Educational Technology, 26*(1), 1–14.

Gaughan, J. E. (2014). The flipped classroom in world history. *History Teacher, 47*(2), 221–244.

Gormely, K., & McDermott, P. (2011). Do you Jing? How screencasting can enrich classroom teaching and learning. *Language and Literacy Spectrum, 21*(1), 12–20.

Graham, G., Parker, S., Wilkins, J. L., Fraser, R., Westfall, S., & Tembo, M. (2002). The effects of high-stakes testing on elementary school art, music, and physical education. *Journal of Physical Education, Recreation & Dance, 73*(8), 51–54.

Herro, D., Kiger, D., & Owens, C. (2013). Mobile technology: Case-based suggestions for classroom integration and teacher educators. *Journal of Digital Learning in Teacher Education, 30*(1), 30–40.

Hwang, G. J., Lai, C. L., & Wang, S. Y. (2015). Seamless flipped learning: A mobile technology-enhanced flipped classroom with effective learning strategies. *Journal of Computers in Education, 2*(4), 449–473.

Kiger, D., Herro, D., & Prunty, D. (2012). Examining the influence of a mobile learning intervention on third grade math achievement. *Journal of Research on Technology in Education, 45*(1), 61–82.

Lei, J. (2009). Digital natives as pre-service teachers: What technology preparation is needed? *Journal of Computing in Teacher Education, 25*(3), 87–97.

Leinonen, T., Keune, A., Veermans, M., & Toikkanen, T. (2014). Mobile apps for reflection in learning: A design research in K–12 education. *British Journal of Educational Technology, 47*(1), 184–202.

Lenhart, A., Ling, R., Campbell, S., & Purcell, K. (2010). *Teens and mobile phones.* Retrieved from http://www.pewInternet.org/2010/04/20/teens-and-mobile-phones/

National Coalition for Core Arts Standards. (2014). National Core Arts Standards: Visual arts model cornerstone assessments. Retrieved from http://www.nationalartsstandards.org/mca/visual-arts

Ng, W. (2015). *New digital technology in education: Conceptualizing professional learning for educators*. New York, NY: Springer.

Norris, C., & Soloway, E. (2008). Getting mobile: Handheld computers bring K12 classrooms into the 21st century. *District Administration Magazine*. Retrieved from http://www.districtadministration.com/article/handhelds-getting-mobile

Prince, M. (2004). Does active learning work? A review of the research. *Journal of Engineering Education*, *93*(3), 223–231.

Roberts, K. A. (2015). *In-class use of cellphones for flipped learning in a high-school art class*. Retrieved from https://sites.google.com/site/kyleroberts0844/professional-work-1

Squire, K., & Dikkers, S. (2012). Amplifications of learning: Use of mobile media devices among youth. *Convergence: The International Journal of Research Into New Media Technologies*, *18*(4), 445–464.

Strayer, J. F. (2012). How learning in an inverted classroom influences cooperation, innovation and task orientation. *Learning Environments Research*, *15*(2), 171–193.

Swan, K., Hooft, M. V. T., Kratcoski, A., & Unger, D. (2005). Uses and effects of mobile computing devices in K–8 classrooms. *Journal of Research on Technology in Education*, *38*(1), 99–112.

Sweeny, R. W. (2008). Unthinkable complexity: Art education in networked times. In M.Alexenberg (Ed.), *Educating artists for the future: Learning at the intersections of art, science, technology and culture* (pp. 85–102). Chicago, IL: Intellect.

Thomas, M. K., Singh, P., Klopfenstein, K., & Henry, T. (2013). Access to high school arts education: Why student participation matters as much as course availability. *Education Policy Analysis Archives*, *21*(83).

Tune, J. D., Sturek, M., & Basile, D. P. (2013). Flipped classroom model improves graduate student performance in cardiovascular, respiratory, and renal physiology. *Advances in Physiology Education*, *37*(4), 316–320.

CHAPTER 12

YOUR IPHONE CANNOT ESCAPE HISTORY, AND NEITHER CAN YOU

Self-Reflexive Design for a Mobile History Learning Game

Owen Gottlieb
Rochester Institute of Technology

Fifth- to seventh-grade players of the mobile augmented reality game (ARG) *Jewish Time Jump: New York* (Gottlieb & Ash, 2013) take on the role of time-traveling reporters for the fictional *Jewish Time Jump Gazette*. Their editor has tasked them with retrieving a story "lost to time." Players, on site in Washington Square Park, in Greenwich Village, New York "land" on site on the eve of the Uprising of 20,000 in 1909, the largest women-led strike in U.S. history. Using their iPhone or iPad, players interact with dozens of digital characters including striking workers, factory owners, labor organizers, journalists, Tammany Hall police, and thugs for hire. Later, players time jump to 1911 and are on site at the Triangle Shirtwaist Factory fire in which 146 people, mostly young Jewish and Italian women, died. During their travels, players receive photographs from the past

Mobile Learning: Perspectives on Practice and Policy
pp. 247–264

based on their GPS location and they pick up various items into their digital inventories: newspapers of the period (some translated from Yiddish), pamphlets, ephemera, and various primary source documents. Toward the end of the game (**Spoiler Alert**), as they travel back in time to the present, they receive a recent news article revealing that the iOS device on which they have played the game is also the nexus of a labor dispute, as Apple works to address serious grievances on the part of Chinese workers at the Foxconn plant who are building the devices. It turns out that the players have been involved in a labor dispute the whole time and they did not even realize it—by using an iOS device.

GAME DESIGN FOR RESEARCH AND LEARNING

Working with a team of more than 20 engineers, historians, researchers, and educators, I designed *Jewish Time Jump: New York*, as both a learning intervention and a design-based research project. The game was funded by The Covenant Foundation and designed and developed at ConverJent: Jewish Games for Learning, an organization I founded in New York. It was built on the ARIS platform for iOS (iPhone and iPad). My doctoral dissertation comprised design-based research conducted on *Jewish Time Jump: New York*. Design-based research is an iterative, proto-theory-testing approach to developing learning theory and design knowledge. Designer-researchers develop a learning environment or intervention over the course of a number of iterative cycles comprising prototyping, field trial, data gathering, analysis, and return to theory and redesign (see Barab & Squire, 2004; The Design-Based Research Collective, 2003; C. Hoadley, 2004; C. M. Hoadley, C. M. 2004). As I piloted and then iterated the game, I worked to gather data in the field from a series of game days with boys and girls in Jewish supplementary schools (after-school programs). I would bring a design to a game day, gather pre- and postsurvey data, conduct interviews, and record video and audio observation of players. The back-end system of ARIS also gathered data about each player's moves in the game. Between iterations, I analyzed the data and made alterations to the game to work toward moving the learners' experiences closer to the learning goals; specifically, those centered on civic and democratic education goals and history education goals. The outcomes of these inquiries included a series of design techniques for helping players to better understand the points-of-view of characters in the game with whom they disagreed. That specific research is published elsewhere (Gottlieb, 2015).

This chapter focuses on the design approach used in the finale of the game, the self-reflexive moment in which the iOS device itself and the player using it are implicated in the historical moment and theme of the game. I drew from self-reflexive traditions in theater, cinema, and nonmobile games to craft the reveal of the connection between the mobile device and the history that the learners were studying. Through centering on this particular design element, I will demonstrate how self-reflexivity can be deployed in a mobile learning experience to tie history to contemporary concerns. What does it mean to bring self-reflexive techniques to mobile learning? What should we consider as we bring these techniques to bear on the mobile learning environment? How can we take advantage of the affordances of mobile self-reflexivity? I explore these questions, and more, through the case of attempting to bring the self-reflexive technique to mobile learning; specifically, in a mobile ARG for teaching history.

Self-Reflexivity as a Means of Tying the Past to the Immediate Present

One immediate way of engaging learners in the study of history is to tie the medium of communication for the lesson itself directly to the enduring historical themes of the subject matter being covered.[1] For issues of history involving economics and social systems, turning to the means of production can be particularly effective at implicating, and therefore involving, the learner in the historical theme. By "implication," I mean that the learners consider their own role in relationship to the enduring historical themes being covered. Perhaps they ask: "what role am I playing in this history as it happens?" The mobile device lends itself to such historical self-reflexivity, or self-reference, for a number of reasons.

The mobile device is ever-present as adults and children consistently turn to their mobile screens throughout the day. For many, smartphones are integrated into our daily living and the way we relate to one another. They are part of our interactions, whether we use them to fact-check or find a restaurant. Their presence and intimate inclusion in our lives provides numerous opportunities and touchpoints through which considering the device itself can be relevant. Used in a history lesson, the mobile device can provide digital primary and secondary sources far more easily than paper, and they can be used on-site, in the field, and outside of the classroom.

If we, as history educators, can move one step further, and tie the mobile medium, the platform of daily digital interactivity of our learners, directly to the historical theme or domain that we are exploring, we may find a different kind of line into student engagement. In such situations, we can ask "what does your smartphone tell you about your role in history?"

This self-reflexive step can provide a different set of opportunities to designers of mobile learning environments and the educators who use mobile history ARGs. As the mobile game hails or interpellates the player (see Althusser, 2006), the self-reflexive technique draws attention to the mobile form itself, as well as the players' involvement with and use of that form.

In this chapter, I argue for the use of self-reflexivity in mobile history games through the use of the case study of *Jewish Time Jump: New York*. I draw from history education theory, civic and democratic education theory, and media theory including cinematic and game studies work. I explore techniques through which designers may avoid pitfalls they are likely to encounter while developing mobile ARGs, and I provide a framework by which the self-reflexive interactive design can work in broader application in technology and learning.

Three key questions arise out of the self-reflexive "turn" that can provide an opportunity for learners to engage in history as active citizens. First, how is this mobile device made and second, what economic and social systems play a role in the processes of putting the device into the hand of a learner/player? The third, and critical, question is: what role have I played, or do I play, in this process? Beyond contemporary mobile games, these provocations driven by a self-reflexive technique can be applied to any new media-based or interactive learning environment.

History: Preparing Engaged Citizens With Enduring Themes

Before addressing the question of how might we move mobile history games closer, as a genre, to their learning goals, it is important to first ask: "What is the purpose of history education?" (Gottlieb, 2015). Barton and Levstik (2004/2012) articulate a particular approach to the goals of history education in their book *Teaching History for the Common Good*. For these authors, history education is to prepare students for participation in a pluralist democracy and to help engage citizens in "collaboration toward a common good" (Preface, paragraph 4). Here, history education is understood as a critical part of civic education. It is about helping learners to participate in a democracy through the ability to engage with political institutions and through participation in civic society (such as through religious institutions, charities, and unions).

Barton and Levstik (2004/2012) cite Turner and Dewey and note that history's goal is to better understand the present. Thus, addressing significant and "enduring themes" or "enduring debates," is essential. An "enduring" theme must speak to the importance of connecting the historic to the present day. In addition, "enduring" implies that better understanding the present day requires exploration and investigation of history (Chapter 4, para. 3). Enduring themes are more than just a connection between present and past. For Barton and Levstik, these themes can teach what it means to participate in a democracy.

Drawing on their own research; as well as research from scholars such as Walter Parker, Benjamin Barber, Shirley Engle, and Anna Ochoa; Barton and Levstik (2004/2012) note the importance of reaching beyond history education that is too often focused on patriotism. Patriotism is also highly valued in dictatorships; but, for a democracy, they argue, we must reach further. Participatory democracy for Barton and Levstik means that people "bring their differences into the open, discuss them, and move forward with mutually acceptable actions where possible" (Chapter 2, Public Education and Democratic Citizenship, para 12.).

How might educators find themes reaching beyond notions of patriotism and toward issues of participation in political institutions and civic society? They can do this by focusing more on the efforts of diverse groups of Americans to achieve full participation in the democracy through themes such as women's suffrage, emancipation, the labor movement, Jim Crow laws, and The Bill of Rights, and less on Thanksgiving and Betsy Ross. Educators can do this, Barton and Levstik (2004/2012) argue, because "Americans will never fully share a national identity so long as their earliest history lessons extol European conquest and praise slaveowners" (Chapter 3, Identification, Participation, and Pluralism, paragraph 12).

If the purpose of history is to prepare learners for citizenship in a pluralist democracy, how might we understand the term "citizen?" Westheimer and Kahne (2004) provide a particularly helpful framework in the case of designing a game-based environment, when they characterize the actions of different kinds of citizens. Their framework includes three models of citizenship: personally responsible, participatory, and justice oriented. The personally responsible citizen obeys laws, works and pays taxes, is responsible in her/his community, and volunteers. To "solve social problems and improve society," one must be personally responsible and be law abiding (p. 240). The participatory citizen knows how government works, organizes community care efforts, and understands strategies for collectively working on tasks. In this model, solving social problems and improving society is based on citizens' active participation, including taking leadership roles within already established community systems and structures. Justice-oriented citizens critically evaluate economic, political, and social structures, looking for root causes. They address injustices and understand democratic social movements, collective and community-based action, and the process of making systemic change. In this model, solving social problems and improving society requires debate and the changing of systems that reproduce patterns of injustice. Westheimer and Kahne point out that a "personally responsible citizen" is not necessarily democratic, as totalitarian regimes also expect their citizens to work, pay taxes, and volunteer. Thus, the authors concentrate on differences in educating the participatory citizen and the justice-oriented citizen.

The authors point out the characteristics of the justice-oriented class that best demonstrated critical analysis and social critique. An educational programmer at a school they highlight, Bayside, described a good citizen in this way:

> A good citizen actively organizes with other people [to address] causes of injustice and suffering.... A good citizen understands the complexities of social issues, political issues, and economic issues, and how they are tied together, and is not always willing to accept the definition of a problem as presented to them by politicians. (Westheimer & Kahne, 2004, p. 255)

Students at this school investigated a number of topics, including child labor practices, and eventually asked school officials to be aware of the practices of the companies from which the school purchased school T-shirts and uniforms. The students in Bayside were more likely to seek out root causes and be more interested in politics than the students in a more participatory citizenship-oriented program. They turned their sites to deep, critical questions. Through the design of *Jewish Time Jump: New York*, I sought to answer questions such as how can those designing mobile history games design promote an understanding of enduring historical themes, linking the past to the present and encouraging active democratic citizenship? How can mobile designs promote participatory and justice-oriented citizenship and learners who have the ability to make critical analyses?

Every medium provides unique affordances, or particular actions that medium facilitates or allows. Mobile devices have delivered particular affordances to teaching and learning, including the ability to use toolsets outdoors and on location. Consider the compass app for navigation, delivery of relevant media such as digital versions of primary historical sources, and recording apps for conducting interviews in the field. Just as mobile affordances have allowed for environmental science investigation on location (Klopfer & Squire, 2004, 2008), they have likewise opened up possibilities for place-based, on-location history gaming, situated documentaries (Gottlieb, 2014, Gottlieb; 2015; Gottlieb, Mathews, Schrier, & Sly, 2014; Hollerer, Pavlik, & Feinman, 1999; Mathews & Squire, 2010; Schrier, 2005; Waglar & Mathews, 2012). It is important that designers and educators pay attention to the unique affordances of the medium in which they are working, especially as those affordances pertain to the subject matter at hand. Here, I wish to suggest another approach to using the mobile device for history education, which is linked not only to the device's mobility, but also linked to the materiality of the device and the attendant history of that materiality—namely, a particular kind of self-reflexive turn: turning the player/learner's attention toward the means of production of the mobile device itself and the how those who use the device are implicated in those means of production.

Self-Reflexivity, Political and Civic Engagement, and Design Choices

Dune (2014) and Kirkland (2007, 2010) have examined self-reflexive techniques in video games, drawing on the Brechtian distancing effect. Dramatist Bertolt Brecht's concept of *Verfremdungseffekt,* or alienation, refers to a designed dramatic effect of foregrounding the form of a performance to refocus the audience/observer attention away from the content and onto that form, the formal devices deployed, and the themes (Dune, 2014, p. 80). Examples of self-reflexive turns in video games, which may or may not be used in a Brechtian sense, might include references to other video games inside of a videogame narrative, or faked "bugs" inside of a game.

In order to push beyond the simple formal use of self-reflexivity in media and toward history, civic, and democratic education; I wish to turn back to Polan (1974) for analysis of Brecht's political and social engagement through art. Polan clearly draws a distinction between formal self-reflexivity and political self-reflexivity in art. He argues:

> conscious and deliberate self-reflexivity—may be nothing other than an expansion and making manifest of inherent qualities of art.... As long as such an art does not connect its formal subversion to an analysis of social situations, such art becomes little more than a further example of the disturbances that go on as we live through a day. (paragraphs 21, 24)

Further, Polan explains Brecht's attitudinal shift in the political. For Brecht, the attitudinal position of the viewing subject springs from an attitudinal position in the work—the political artwork embodies a difference between the way things are and the way they can be. "Political art compares an image of human beings as 'unalterable' to one of them as 'alterable and able to alter' (quoted from 'The Modern Theatre Is the Epic Theatre')" (paragraph 28). Following Polan's analysis, mobile designs for fostering civic engagement must move beyond self-reflexivity focused only on consciousness of the medium and toward consciousness of a citizen's role in effecting social change, such as change in a democracy.

This understanding of self-reflexivity in the service of suggesting potential civic action and change-making leads to the design rationale for self-reflexivity in *Jewish Time Jump: New York*. The rationale for these design choices draws on narrative techniques used previously in video games and cinema. Here I am referring to a narrative device of emotional, surprising, late-stage, self-reflexive reveals that implicates the player or audience member's own ethical choices—making the player or audience member complicit in the relevant action of critique. In games, this is exemplified by Warren Spector's videogame *Deus Ex* (Spector & Smith, 2000, as discussed by Gee, 2007) in which the player's complicity in actions they likely would not have sanctioned is revealed after the fact. It is also exemplified in Brenda Brathwaite's board game and meditation on the Holocaust, *Train* (2009). In *Train*, players work to move trains with cargo to their station, only to later discover where the trains are headed—to concentration camps. The realization as to where the trains are heading usually concludes the game (Brathwaite & Sharp, 2010; Brophy-Warren, 2009). Among many themes, the game is primarily about complicity. That complicity is delivered through the procedural rhetorics[2] of unaware action or action-in-denial and its potentially horrific, even evil, consequences.

Cinema is an important reference for modern mobile media self-reflexivity because filmmakers and video artists have had a long head start on innovating the self-reflexive turn in audio-visual media. The self-reflexive and political turn in cinema is exemplified by Vertov's (1929/2002) *Man With a Movie Camera*, which showed the modes of production of the film itself by showing the acts of cinematography and editing in its creation. Yet it is Block's (1973) short film *No Lies* that broke ground in terms of ethically implicating the film viewer in fraught acts of viewing and intrusion. This is because the viewing of *No Lies* itself reproduces a violation of its subject (a woman interviewed about an assault) and, as the film itself violates the audience's implicit trust, all is not what it seems. At the conclusion of the documentary, the filmmakers reveals the act of violence was in fact, staged, violating the audience. An understanding of this history of games and films making self-reflexive turns that simultaneously hail the player and audience as complicit and implicated, paves the way for such turns in mobile ARGs and situated documentaries.

SELF-REFLEXIVE DESIGN IN THE FIELD: *JEWISH TIME JUMP: NEW YORK*

Initial Design Drawn From Theory

From the earliest pilot of *Jewish Time Jump: New York*, a design concern was to powerfully tie the history of the Uprising of 20,000 and the Triangle Shirtwaist Factory fire to the present day. I theorized, following the lead of the aforementioned models, that using a narrative self-reflexive turn to the mobile device itself would tie the historic moments to the immediate present through enduring themes and controversies. In addition, such a turn should, in theory, implicate the player personally in the historic moments, as the themes of labor disputes, employment, and production shift from the historic shirtwaist workers and factory owners in 1909–1911 into the present, the immediate, and the global. By linking the present and past through enduring themes in this way, and by involving the learners in a personalized way in the gameplay, I was reaching for pathways toward participatory and justice-oriented citizenship engagement. The method of making a personal connection for the player was originally envisioned in the finale of the game, in which it is revealed to the players that the mobile device upon which they have played the game is itself the nexus of labor disputes.

Upon returning from their reporting and investigating adventures in the early 1900s to "present day," players then have present-day encounters. These encounters include videos of a contemporary investment banker commenting on modern labor relations and a rabbi telling the story of her great grandmother who survived the Triangle Shirtwaist Factory, went on to testify in court against the owners and sent on to fight for fire-code regulations. In addition, players receive a contemporary news story discussing the workers at Foxconn in China who build iPhones and iPads, the recent disputes over worker mistreatment and unhealthy conditions, and Apple's moves to improve conditions (Reuters, 2012).

This end-game reveal was consciously paralleled with the narrative of the player's return through time travel from the past to the present day. The decision to use the self-referential turn on the device itself was intended to transcend the diegesis of play and to raise to consciousness the medium of the mobile device itself. This design decision is not intended for an end result of the breaking of the "fourth wall," creating distance between the player and game, but rather to deepen the player's personal play experience, to implicate the player in history, and to tie the action of the player in playing the game directly into the enduring themes of history.

The players' use of the game device itself also has implications and is imbricated within the power structures and struggles the player investigates and explores in the other content of the game. Just as the player meets strikers, organizers, and bosses from 1909–1911, it is clearly communicated that: (1) there are still labor clashes today and (2) that the players themselves are involved to some degree with those struggles; the device they are holding (and usually the device their family owns) is the subject of labor disputes. The device the learner is carrying may have been built in China by people working in conditions that would not be acceptable in the United States and that have been blamed as the cause of the workers' deaths.

Such a moment, developed in the pilot, I theorized, should raise questions and provide educators the opportunity to raise their own questions, such as: How is one implicated or active in history, even when unaware? How can we pay attention to where our products come from and how they are produced? The need to implicate the player in the enduring theme in history, in this case labor disputes, was a clear design goal. The intention was that a player should not finish playing the game without being confronted by such questions of player implication being raised within the game.

Learning From Implementation Challenges

Jewish Time Jump: New York contains more than 20 digital characters and dozens of events and historic artifacts. Thus, the actual process of *deploying* the self-reflexive and player-implicating game event in a sprawling, mobile, place-based game proved to be complicated. My hope is that through sharing the challenges and solutions in the context of developing and testing mobile place-based augmented reality gaming that other designer-educators will be able to benefit from what I learned, and they will have an advantage in designing other mobile self-reflexive calls for citizen action.

The study was conducted over the course of a number of game days with analysis and game iteration in between those events. Forty-three fifth- to seventh-grade learners (24 girls, 21 boys) from supplementary schools participated in gameplay over the course of six game events and three design iterations (some game days occurred within a few days of each other to allow for more players in each play session, or, in one case, close single-player observation). During the game days, the learners took presurveys, were then trained on the device, and were set into the park to play the game. Following each play session, they took a postsurvey and participated in debriefing and interview/discussion. Some players also took part in longer interviews. In between the game events, I analyzed the data, including video observation footage and server logs that recorded each player's specific moves. This chapter can only address the self-reflexive design technique (for full data sets and detailed description of the full study see Gottlieb, 2015).

In the first event, the pilot of the game, given the sprawling nature of the narrative and the ability for learners to immerse in various historical artifacts, only a few students completed the game, and therefore did not receive the Foxconn news story. In the first iteration, I tightened the game structure to reduce play time, but in addition, I added an *Afikoman*, referring to a hidden bonus or special event, based on the idea of the "Easter egg" in video games (hidden bonus material or levels). The Afikoman refers to a game played during the Passover seder in which a piece of matzah is hidden, and the seder may not be complete until it is found or returned by the children tasked with finding it. The digital Afikoman held both contemporary videos and the news article. In this way, all players who did not reach the conclusion of the game would be granted the Afikoman by the educators at the end of play so that they could experience them as well. In order for the self-reflexive late stage reveal to work, the players must encounter it. In a place-based game with multiple routes and possible player paces, that means accounting for providing the reveal in a nonlinear fashion.

In field testing this next iteration, I found that some students picked up, viewed, and did not read the article about Foxconn, but rather jumped ahead to complete the game. Even when presented with a second chance to read and review the article along with students who received the Afikoman, one girl's short answer in the postgame survey reveals a lack of connection to the enduring historical themes:

> [Crossed out]: I was kind of rushing to finish, but I think it had to do with the difference between Jewish rights today and 100 years ago [end cross out]. Jewish rights today are very different from 100 years ago. Human rights in general were different! Some people were so high up the hill, they thought they had more rights than other people! Bosses treated workers like scum and would do anything to earn money, even risk lives of workers!

It is possible the learner is not connecting the global present-day labor issues with those of the past. Regarding the humane treatment of workers, she writes, "human rights in general were different [than they were 100 years ago]." It is also possible that the article regarding Apple working to improve its conditions provided a contrast to owners' disregard for worker safety in the years prior to the post-Triangle fire labor laws. Given this learner's dual opportunity for exposure to the article, both during gameplay and again during the Afikoman following main play, the persistent lack of connection to the enduring historical theme signals the need for a better design solution to reach the educational goal.

In the next iteration, I took a different approach, based on Levstik and Barton's (2005/2010) alternative strategies for teaching history of "start locally, connect globally," and "start globally, connect locally" (Chapter 6, Sections 2 and 3). For example, shall a teacher introduce the themes with examples from the town the learner lives in, or with related events in a location thousands of miles away? In this case, I tried what I call a suprareveal. Rather than giving away the reveal at the end (or in the case of the Afikoman, near end) of the game, which was a local to a global strategy, I added a global starting point to frame the entire game. At the beginning of the game, the learners receive an article in the form of a clue prior to meeting their editor. It arrives in the form of a 2013 article about a recent garment factory collapse in Bangladesh and the resulting 1,100 deaths (Yardley, 2013). Learners are guided to read just the first two paragraphs of the article and then move on to find their editor.

In field testing this version of the game, the discussions of the factory collapse began immediately as players asked how this related to the game and their mission (they had watched an initial introductory video clip telling them to go to the park to meet their editor, but had not yet departed from the staging area to the park). Questions included "What does this article have to do with Jewish history? and "What is the date on the article?" Although surveys in previous iterations revealed lack of awareness of contemporary global issues of worker safety and treatment, in this latest iteration, there was no evidence in either postgame interviews or open-ended answers on postgame surveys of lack of awareness of the contemporary global issues of worker safety. Neither was there evidence of such a confusion or misunderstanding in the video footage in the Afikoman about Foxconn.

Learners read the Bangladesh article out loud prior to the game and asked questions about the article, which provided evidence of awareness of the historic theme. There was also a shift from evidence of lack of awareness regarding contemporary labor struggles, such as the girl quoted above, to an iteration devoid of disconfirming cases of awareness. Given both of these observations, it appears that the use of a suprareveal prior to or at the early stages of play in a mobile ARG can offset issues of lack of attention to contemporary cases as well as provide a conscious frame for the game to come. In addition, the suprareveal allowed for maintaining the surprise reveal either upon reaching the conclusion of the game, or in receiving the Afikoman. In other words, by providing the Bangladesh article early in play, I could set up the historic theme while maintaining the late-stage reveal of the Foxconn article. The player could be primed with the theme and still discover that they had been unaware that, for the length of the game, they had been using a device that is at the center of labor disputes.

Beyond Iterative Game Design and Into Curriculum

Invoking the self-reflexive turn, just as any provocative learning moment generated in a game-based environment, also requires reflection and processing. Once I was assured that players on the ground encountered the self-reflexive reveal (the surprise that the iOS device itself was the subject of labor disputes), and that they could make connections between the past and present events through the enduring historic themes, I took on an additional project to extend discussion of contemporary labor practices. While the formal research study of the design-based game iterations concluded, I began a collaborative project between my games organization, ConverJent, and the Jewish Women's Archive, funded by The Covenant Foundation, to develop a modular curriculum for the game. This curriculum, made of four modules which can be used together, individually, or in various combinations, can be deployed by educators in various settings, with various time constraints, by using any number of modules covering different, yet related, subjects. In the Contemporary Labor Issues module of the curriculum, we included guiding questions about the *Reuters/New York Times* article about Foxconn as well as the Bangladeshi factory collapse. These included the following questions:

- What situation is this article describing? What are the issues that factory owners and workers need to address?
- How is this story similar to, or different from, labor issues in New York City in the early 20th century?
- As American citizens, do we have a responsibility to change labor practices in other countries? Why or why not? (ConverJent and Jewish Women's Archive, 2015)

While other aspects of the Contemporary Labor Issues module address contemporary labor issues in America, one area that they do not currently address is contemporary sweatshop conditions in New York City. Were there further iterations, the inclusion of questions and issues around sweatshops in New York City would be an important addition to focus on continuing opportunities for civic engagement on the site of the game.

CONCLUSIONS

Self-reflexive techniques have a long history in other media. Among a number of goals for the wider research project, I sought to explore the affordances of self-reflexive techniques in mobile history learning. How could self-reflexive techniques provide learning opportunities that would be bolstered by the devices used to play the game? By pursuing deepened conversations about the contemporary labor practices, I aimed to leverage self-reflexivity in *Jewish Time Jump: New York* to help learners understand their personal connections to, and implication in, historical themes. I also worked to provide opportunities for discussion to lead to enhanced participatory and justice-oriented citizenship. Through the historic figures in the game, including labor organizers and manufacturers who organized around their own interests, the game emphasizes issues-based advocacy and the ways in which citizens organize. By positioning players in the role of journalists, the game suggests the connections between citizen journalists, who ask challenging questions and seek multiple sources, and historians who use similar approaches. Through the game narrative, players also encounter the alignment of political machinery (Tammany Hall) and business leaders. Through all these themes, the game works to show power structures and how civic engagement and civil institutions can make change, whether that is the end of the 1909 Uprising which led to all but a handful of garment shops becoming unionized (the Triangle Factory was one of the few that did not), or whether that meant that some of the contemporary local newspapers supported the manufacturers (*The New York Times*) while others supported the strikers (*The New York Call*, *The Daily Forward*). The self-reflexive turn frames all the various themes and questions with the broader question: how are you, the player, involved in history when you use a mobile device?

While longitudinal studies would be required to determine whether players took on new participatory or justice-oriented civic engagement activities; the Foxconn and Bangladesh articles, when used together, generated discussion of contemporary labor disputes in the context of historic labor disputes. Those two articles, used in concert, also eliminated confusion and erroneous conclusions about contemporary labor disputes. This method moved learners closer to the civically engaged model of self-reflexivity that Polan (1974) points out in Brecht's work. Brecht's desire for social and political engagement aligns, in many ways, with the models of participatory and justice-oriented citizenship education, and so it is fitting that his formal techniques can lend assistance to design for mobile history learning. By building curriculum around the questions raised through self-reflexivity, educators can further guide discussion, once the provocation in the game has taken place.

What additional significance can self-reflexivity hold for mobile learning designers and educators, or future interactive media designers and educators? First, the ability to implicate a learner/player in the content through the form of the medium means that individual learners need not arrive on the scene already having a personal connection to the content, but only the form—the use of the mobile device, or other technology platform being used. The form connection is far more likely given that the learner/player must use the form to reach the content of the game or simulation, when there is no guarantee that the learner has had previous exposure to the content material. You cannot read this sentence without engaging in the process of textual production and reception. Likewise, the mobile user is implicated in mobile manufacturing, design, sales, and marketing; the virtual reality user is implicated in the means of production of the headsets, vests, visors and other devices. If our learning content can be connected to history and economics, then so can our relationship to the medium through which we teach.

And yet, the implication of the player/user in historic themes requires further design and contextualization. The design must ensure the player's experience of the self-reflexive turn—something not so simple in a mobile game. The experience must be adequately prepared so that the notion of complicity occurs through participation in the process—a late-stage reveal. Complicity, or implication, requires participation in the text (here text refers to any piece of media) first. The frame by which to understand that context might need to be primed early, such as through foreshadowing or a suprareveal in order to clearly establish the enduring historic theme. Ideally, the implication of the player/user is also embedded in reflective discussion in a curriculum so that it can be unpacked and digested. Then educators can link Brechtian engagement to civic engagement—to work within a democracy to shift from the way things are to "the way things could be." Neither iteratively designing a game nor taking time unpacking the self-reflexive turn with learners are easy processes, yet, self-reflexive techniques for mobile media, augmented reality, virtual reality, and future forms of media mean that designers have a powerful mode for turning the medium back on itself in order to impart the player/user's position in history as an active participant and agent. With self-reflexivity, mobile devices can become mirrors.

NOTES

1. Applying the self-reflexive technique here, to this very chapter, we can consider the historical subject matter and theme at hand, for example, labor history, and then inquire: What were the means of production of the text you are reading right now? How was it written, reviewed, and edited? Were the editors or authors paid? Is it a part of a different kind of economic system? How was the text published and distributed so that you are now reading it? Are you reading on a digital reader, in a paper bound copy? What role do you play as reader, consumer, or reviewer of this text? How does history play into the production of this text?
2. Procedural rhetorics (Bogost, 2007) refers to persuasion and argumentation in games constructed through the modeling of systems and procedures. As games are a procedural medium (players follow action paths and

procedures through a system of rules), Bogost argues that the manner in which the systems are modeled creates a rhetoric.

REFERENCES

Althusser, L. (2006). Ideology and ideological state apparatuses (notes towards an investigation). *The Anthropology of the State: A Reader, 9*(1), 86–98.

Barab, S., & Squire, K. (2004). Introduction: Design-based research: Putting a stake in the ground. *The Journal of the Learning Sciences, 13*(1), 1–14.

Barton, K. C., & Levstik, L. S. (2012). *Teaching history for the common good* [Kindle Keyboard version]. Retrieved from Amazon.com. (Original work published in 2004)

Block, M. (Producer & Director). (1973). *No lies* [Short film]. United States: Direct Cinema, Ltd.

Bogost, I. (2007). *Persuasive games: The expressive power of videogames*. Cambridge, MA: The MIT Press.

Brathwaite, B. (2009). Train [Board game]. Self-published.

Brathwaite, B., & Sharp, J. (2010). The mechanic is the message: A postmortem in progress. In K. Schrier & D. Gibson (Eds.), *Ethics and game design: Teaching values through play* (pp. 311–329). Hershey, PA: IGI Global.

Brophy-Warren, J. (2009, June 24). The board game no one wants to play more than once. Retrieved from http://blogs.wsj.com/speakeasy/2009/06/24/can-you-make-a-board-game-about-the-holocaust-meet-train/

ConverJent and Jewish Women's Archive. (2015). *Jewish Time Jump: New York*: Lesson plans about Judaism and labor to accompany the mobile game (Curriculum for *Jewish Time Jump: New York*). Contemporary Labor Issues, Module #4. Retrieved from http://jwa.org/teach/jewishtimejump

The Design-Based Research Collective. (2003). Design-based research: An emerging paradigm for educational inquiry. *Educational Researcher, 32*(1), 5 –8. https://doi.org/10.3102/0013189X032001005

Dunne, D. J. (2014). Brechtian alienation in videogames. *Press Start, 1*(1), 79–99.

Gee, J. P. (2003). *What video games have to teach us about learning and literacy* (1st ed.). New York, NY: Palgrave Macmillan.

Gottlieb, O. (2014). *Jewish Time Jump: New York*. In K. Schrier (Ed.), *Learning, education, & games* (Vol. 1, pp. 210–211). Pittsburgh, PA: ETC Press. Retrieved from http://press.etc.cmu.edu/content/learning-education-and-games-volume-one-curricular-and-design-considerations

Gottlieb, O. (2015). *Mobile, location-based game design for teaching Jewish history: A design-based research study* (Doctoral dissertation). New York University, New York.

Gottlieb, O. (2017). New design principles for mobile games history games. *GLS 12 (Games Learning Society) Proceedings*. Pittsburgh, PA: ETC Press.

Gottlieb, O., & Ash, J. (2013) *Jewish Time Jump: New York* [Video Game]. New York, NY: ConverJent.

Gottlieb, O., Mathews, J., Schrier, K., & Sly, J. (2014). Mobile history games: Challenges, frameworks, and design principles. *Proceedings GLS 10 Games + Learning + Society Conference*. Madison, WI: ETC.

Hoadley, C. (2004). Fostering productive collaboration offline and online: Learning from each other. In M. Linn, E. A. Davis, & P. Bell (Eds.), *Internet environments for science education* (pp. 145–174). Mahwah, NJ: Erlbaum.

Hoadley, C. M. (2004). Methodological alignment in design-based research. *Educational Psychologist, 39*, 203–212. http://doi.org/10.1207/s15326985ep3904_2

Hollerer, T., Feiner, S., & Pavlik, J. (1999). Situated documentaries: Embedding multimedia presentations in the real world. In *The Third International Symposium on Wearable Computers, 1999. Digest of Papers* (pp. 79–86). http://doi.org/10.1109/ISWC.1999.806664

Kirkland, E. (2007). The self-reflexive funhouse of Silent Hill. *Convergence: The International Journal of Research into New Media Technologies, 13*(4), 403–415.

Kirkland, E. (2010). Discursively constructing the art of Silent Hill. *Games and Culture*. Retrieved from http://gac.sagepub.com/content/early/2010/05/11/1555412010364976.abstract

Klopfer, E., & Squire, K. (2004). Getting your socks wet: Augmented reality environmental science. In *Proceedings of the 6th international conference on Learning Sciences* (pp. 614–614). Retrieved from http://dl.acm.org/citation.cfm?id=1149126.1149238

Klopfer, E., & Squire, K. (2008). Environmental detectives—The development of an augmented reality platform for environmental simulations. *Educational Technology Research & Development, 56*(2), 203–228. https://doi.org/10.1007/s11423-007-9037-6

Levstik, L. S., & Barton, K. C. (2010). *Doing history: Investigating with children in elementary and middle schools* (4th ed.) [Kindle Keyboard version]. Retrieved from Amazon.com. (Original work published in 2005)

Mathews, J. M., & Squire, K. D. (2010). "Augmented reality gaming and game design as a new literacy practice." In K. Tyner (Ed.), *Media literacy: New agendas in communication* (pp. 209–232). Austin, TX: University of Texas at Austin.

Polan, D. B. (1974). Brecht and the politics of self-reflexive cinema. *Jump Cut, 1*. Retrieved from http://www.ejumpcut.org/archive/onlinessays/JC17folder/BrechtPolan.html

Reuters. (2012, August 22). Auditors find improved working conditions at an Apple supplier. *The New York Times*. Retrieved from http://www.nytimes.com/2012/08/23/technology/23iht-apple23.html

Schrier, K. (2005, September). *Revolutionizing history education: Using augmented reality games to teach histories* (Master of Science, Comparative Media Studies). MIT.

Schrier, K. (Ed.). (2014). Designing digital games to teach history. In *Learning, education and games* (pp. 73–91). Pittsburgh, PA: ETC Press. Retrieved from http://dl.acm.org/citation.cfm?id=2811147.2811152

Spector, W., & Smith, H. (2000). *Deus Ex* [Video game]. Austin, TX: Eidos Interactive.

Vertov, D. (Director). (2002). *Man with a movie camera* [Documentary film]. Soviet Union: Image Entertainment. (Original work published in 1929)

Wagler, M., & Mathews, J. (2012). Up river: Place, ethnography, and design in the St. Louis River estuary. In S. Dikkers, J. Martin, & B. Coulter (Eds.), *Mobile media learning: Amazing uses of mobile devices for learning* (pp. 41–62). Lulu.com.

Westheimer, J., & Kahne, J. (2004). What kind of citizen? The politics of educating for democracy. *American Educational Research Journal*, *41*(2), 237–269. http://doi.org/10.3102/00028312041002237

Yardley, J. (2013, July 2). After disaster, Bangladesh lags in policing its maze of factories. *The New York Times*. Retrieved from http://www.nytimes.com/2013/07/03/world/asia/bangladeshi-inspectors-struggle-to-avert-a-new-factory-disaster.html

ABOUT THE CONTRIBUTORS

ABOUT THE EDITORS

Danielle Herro is an associate professor of digital media and learning at Clemson University and doctoral coordinator for the Learning Sciences in the College of Education. From 2013–2016 she received an Edmund W. Gordon/MacArthur Foundation Fellowship for 21st Century Learning and Assessment. Dani teaches courses on social media, games, and emerging technologies, recently developing and teaching, "Theoretical Foundations of Games for Learning." At Clemson, she codesigned and opened Digital Media and Learning and Gaming Labs in the School of Education. Dani has published numerous journal articles on the impact of mobile learning in K–12 education including research detailing case-based suggestions for using mobiles in the classroom, bring-your-own-device initiatives, and the impact of mobile learning interventions on math achievement. Her current research interests focus on investigating game based learning in classrooms, the intersection of games, identity and social practices, and the efficacy of teacher professional development towards integrating digital media in science, technology, engineering, art and mathematics.

Sousan Arafeh is an assistant professor and coordinator of doctoral education in the Department of Educational Leadership and Policy Studies at Southern Connecticut State University. Dr. Arafeh has worked in the fields of education, policy, and communications as a classroom teacher and a researcher and has published and lectured nationally and internationally, particularly in the area of science, technology, engineering, and mathematics education. While at the American Institutes for Research, she consulted for key education-related agencies and foundations nationwide including the U.S. Department of Education and the National Science Foundation. Dr. Arafeh served as the deputy director of the National Center for Technology Innovation—a National Center to innovate assistive technology for improved outcomes for students with disabilities. She also participated in the development of two of the Office of Technology's Education Technology Plans and collaborated with the Pew Research Centers to research and coauthor the seminal *Digital Disconnect* report. Currently,

Dr. Arafeh is exploring further ways to advance social justice in formal and informal technology-connected spaces.

Rich Ling (PhD, University of Colorado, sociology) is the Shaw Foundation Professor of Media Technology, at Nanyang Technological University, Singapore. He has an adjunct position at the University of Michigan. For the past 2 decades Ling has studied the social consequences of mobile communication. He has written *The Mobile Connection* (Morgan Kaufmann, 2004), *New Tech, New Ties* (MIT, 2008) and most recently *Taken for Grantedness* (MIT, 2012). He is the editor-in-chief of the *Journal of Computer-Mediated Communication* and is a founding coeditor of *Mobile Media and Communication* (SAGE) and the Oxford University Press Series *Studies in Mobile Communication*. He was recently named a fellow of the International Communication Association.

Chris Holden is an associate professor in the Honors College at the University of New Mexico where he teaches small, interdisciplinary undergraduate courses. He teaches his students math (Chris's PhD was in number theory) and how to analyze and begin to make vernacular modern medialike games. He is also affiliate faculty in educational linguistics and organization, information, and learning sciences at UNM. Broadly, he studies games and learning. He has designed mobile augmented reality (AR) games for use as classroom curricula, museum experiences, community contexts, research diagnostics, and just for fun, in relation to many topics and themes. In particular, his joint work with Julie Sykes on one AR project, Mentira, has inspired a flurry of creative research worldwide, especially in language learning contexts.

ABOUT THE AUTHORS

Jodi Asbell-Clarke has over 20 years designing and researching curriculum and professional development for STEM (science, technology, engineering, and mathematics) teachers. In 2009, she cofounded EdGE at TERC, a team of game designers, educators, and researchers studying implicit STEM learning in digital games. Jodi's background includes an MA in math, an MSc in astrophysics, and a PhD in education. She was an onboard software verification analyst for IBM during the first 25 missions of the Space Shuttle and taught at the laboratory school at University of Illinois.

Helen Crompton, PhD, is an assistant professor of instructional technology at Old Dominion University, Virginia. She gained her PhD in educational technology and mathematics education from the University of North Carolina at Chapel Hill. Her research is focused on mobile learning and the effective integration of technology into K–12 education. Dr. Crompton works as a consultant for two United Nations Agencies (United Nations, Educational, Scientific, and Cultural Organization and International Telecommunication Union) to research, author and edit publications summarizing research on mobile learning. Dr. Crompton is also a faculty member for the International Society for Technology in Education.

Aijuan Cun is a former part-time bilingual teacher in China. Cun received her master degree in TESOL at SUNY Fredonia and is currently a doctoral student in the Curriculum, Instruction, and the Science of Learning program at University at Buffalo (SUNY). Her research interests include digital multimodal composition, literacy practices of immigrant students, and youth community inquiry.

Benjamin Deaton is dean of the Center for Innovation and Digital Learning at Anderson University and was the founding director of the South Carolina Center of Excellence for Mobile Learning. He provides leadership for Anderson's Mobile Learning Initiative that provides all students and faculty members with mobile devices and focuses on the intentional integration of mobiles to support and enhance teaching and learning. His unit is responsible for learning technologies, learning space design, and online and blended learning at Anderson.

Cynthia C. M. Deaton is an associate professor of science education in the College of Education at Clemson University. Her research focuses on science teacher education, reflective practice, mobile learning in science, and professional development. She has facilitated a variety of professional development experiences integrating science, technology, and inquiry for pre-K–12. Currently, she teaches qualitative research and science methods for elementary education.

Judith Dunkerly-Bean, PhD, is an award-winning assistant professor of literacy in the Department of Teaching and Learning at Old Dominion University. She earned her degree at the University of Nevada, Las Vegas. Her work has been published in the *Journal of Literacy Research*, the *Journal of Adolescent and Adult Literacy*, *Comparative Issues in Education* and *Language and Literacy*, as well as in several edited volumes and handbooks. She is the coauthor of *Teaching Young Adult Literature: Developing Students as World Citizens*. Judith's research utilizes transnational feminist and cosmopolitan theory to explore the intersection of critical literacy, social justice, and human rights.

Laura A. Gillespie (University of Baltimore)—after earning a BA and MA in humanities fields—English literature and medieval studies, respectively—and acting as a program assistant for a D.C.-based high school science program that uses national parks as classrooms, Laura earned her ScD in information and interaction design from the University of Baltimore while exploring how technologically mediated experiences might enhance history learning in informal environments such as national historical parks. She is currently an English language copy editor at the University of Maryland's National Foreign Language Center and teaches advanced research writing at the University of Maryland University College.

Abigail Gloss is a former adolescent, adult, and migrant education teacher. Gloss currently works with preservice education students as a visiting assistant professor in the Language, Learning & Leadership department at Fredonia State University of New York. She is also a graduate student in the Curriculum, Instruction, and the Science of Learning program at the University at Buffalo (SUNY).

Owen Gottlieb, PhD, is an assistant professor of interactive games and media at the Rochester Institute of Technology. He founded and directs the Initiative in Religion, Culture, and Policy at the RIT MAGIC Center. Gottlieb's research and game design work spans religious literacy, the acquisition of cultural practices, as well as interactive media for healing. Gottlieb's latest digital game prototype was sponsored by the National Endowment for the Humanities. He is currently principal investigator on a grant from the National Science Foundation. Jewish Time Jump: New York (featured this volume) was nominated for Most Innovative Game by the Games for Change Festival.

Liz Kolb is a clinical assistant professor at The University of Michigan in Ann Arbor, MI. Her expertise is in technology and teacher education. Her studies have focused on student's use of everyday technologies in formal schooling as well as developing frameworks for assessing effective technology use in schools. She authored *Toys to Tools: Connecting Student Cell Phones to Education, Cell Phones in the Classroom: A Practical Guide for the K-12 Educator, Help Your Child Learn With Their Cell Phone,* and *Web 2.0, Learning First, Technology Second*. In addition, Liz has published numerous articles and book chapters on new technologies and education.

Michael A. Kuszpa is a veteran middle school science teacher in New Haven, Connecticut who is currently a pursuing a doctorate in educational leadership at Southern Connecticut State University. Previously he served 9 years in the United States Army to include one combat tour in Afghanistan as a captain in the United States Army. Kuszpa received his MS in science education from Southern Connecticut State University (2012) and a BS in marine science from the University of Miami (1998). In his free time Kuszpa enjoys fishing and boating on Long Island Sound with his family.

James L. Larsen is a cofounder and lead developer at Educational Gaming Environments (EdGE) at TERC. James's work at EdGE focuses on ways to integrate STEM content through project and STEM-infused games, especially as relates to opening the door for learners to get them outside to learn from nature and the real world. He has taught in both public and private schools as a STEM educator as well as created curriculum and professional development for NSTA, Scientific American Frontiers, NOVA and the GLOBE Program among others. He has a bachelor's in zoology and MEd in educational leadership.

Sandra M. Linder is an associate professor of early childhood mathematics education in the College of Education at Clemson University. Her research centers on supporting and sustaining teacher quality in early childhood mathematics through professional development. She has taught in a variety of elementary settings and has worked intensively with in-service and preservice mathematics educators from infant settings through fifth grade.

Kathryn MacCallum is a senior lecturer in computing and coordinator of the Postgraduate Programme at Eastern Institute of Technology, New Zealand. Her research focuses on mobile learning and other emerging technologies in supporting education. Dr. MacCallum has been a guest speaker for many academic events and has authored over 30 academic articles including book chapters, journals, and conferences, mainly in the areas of mobile learning and educational technology. She is an editor for the *Journal of Information Technology Education*, associate editor of the *International Journal of Mobile and Blended Learning*, and a member of several national and international research committees.

Barb MacEachern is the director of outreach for the Educational Gaming Environments group (EdGE) at TERC, focusing on ways to connect EdGE's work to the outside world. This world includes teachers, students, gamers, academics, parents, game designers ... anyone with an interest in how game-based learning can open doors that allow engagement with the science that is all around us. Barb has 20 years of experience in community building, facilitation, social emotional learning, professional development, out-of-school time curriculum design, program implementation, and educational research.

Thomas L. Mitchell is a doctoral student at Southern Connecticut State University in the Educational Leadership and Policy Studies Department. His area of focus is cultural competency and cultural relevance/responsiveness. Concurrently, he serves as an adjunct professor in the School of Education and School of Music at SCSU and as an administrator in the New Haven Public School District. He has presented work on cultural competence at the Northeastern Educational Research Association Conference and has been recognized as a distinguished scholar by the Golden Key International Honour Society.

Char Moffit, PhD, is an assistant professor of literacy education at the California State University, Chico. Dr. Moffit's research interests include integrating science and literacy instruction in the primary grades through project-based learning. She is also interested in mobile technology use as it relates to literacy learning. She teaches courses in children's literature and literacy methods.

Jennifer L. Motter, PhD, has a background in art education, instructional technology, school administration, and digital design. She earned a doctorate from The Pennsylvania State University in 2012. Her doctoral research "Feminist Art Curriculum: Politicizing the Personal via Cyberpost Activism" involves socially responsible and culturally responsive art education. Motter utilizes the empowerment potential of new media to facilitate leadership opportunities for others, including marginalized groups, in order to promote social justice. She is K–12 liaison and past-president of the National Art Education Association Women's Caucus. Motter develops and instructs art and mobile technology courses at a rural secondary school in Western Pennsylvania.

Aysenur Ozyer is a PhD student in information and learning technologies at CU Denver currently working on her doctoral dissertation. She also works as an instructional designer, designing online courses in higher education. Her research interests include instructional designer's professional identity development, professional learning in instructional technology and design, competency based learning, and student engagement in e-learning.

Merve Pamuk is a former geography teacher in Turkey. Pamuk holds a MA in geography information systems from Ege University. She is currently a master's student in the Learning and Instruction Department at the State University of New York at Buffalo. Her main research interest is using geographic information system technology in education, with a special focus on spatial and geographic thinking.

David Parsons is National Postgraduate Director at The Mind Lab by Unitec, responsible for a national program in digital and collaborative learning for in-service teachers. He holds a PhD in information technology and has wide experience of both academia and the IT industry in the United Kingdom and Australasia. He is the founding editor-in-chief of the *International Journal of Mobile and Blended Learning* and author and editor of many books, chapters and articles on various aspects of technology enhanced learning, with a particular interest in mobile learning. He is a longstanding member of the International Association for Mobile Learning.

Judy Perry oversees design, development, and research for several MIT STEP Lab projects involving games and simulations, and their integration into formal or informal learning settings. Perry's research interests include location-based games, ubiquitous "casual" games, mobile games, and digital-materials designed to foster engagement with STEM. Judy also leads professional development training workshops and enjoys close collaborations with partner institutions (including zoos, nature centers, libraries, science museums, living history museums, out-of-school programs and K–12 classrooms). Perry holds a BA in American studies from Yale University, and an EdM from the Harvard Graduate School of Education.

Ryan M. Rish is a former high school English teacher and is currently an assistant professor in the Department of Learning and Instruction at University at Buffalo (SUNY). Rish earned his doctorate in adolescent, post-secondary, and community literacies from The Ohio State University. Rish's research interests involve the literacy practices of adolescents, especially in regard to how those practices relate to the institutions, social spaces, and geographic places in which they are enacted.

Elizabeth Rowe, a senior researcher at TERC, studies and develops innovative uses of technology in and out of schools with a current focus on game-based learning. Dr. Rowe is currently the director of research for the Educational Gaming Environments group at TERC where she oversees the research design, data collection, and analysis for all EdGE projects. Elizabeth's background includes a bachelor's in math and a PhD in human development and family studies.

Kyle A. Roberts has been teaching art to high school students for 7 years, the last 3 at Cherokee Trail High School in Colorado. A graduate from Rocky Mountain College of Art and Design, Kyle is now completing his MA in information and learning technologies from the University of Colorado Denver. With experience teaching in a variety of mediums, he currently teaches drawing and painting, and visual art in the school's International Baccalaureate program. Kyle's students have shown their work in the Denver Art Museum, Pratt University NYC, been honored in Carnegie Hall NYC and exhibited at the Museum of Modern Art in Denver. His teaching centers on fostering critical thinking, creativity and developing technical mastery.

Sara Vogel is a doctoral student in urban education at the Graduate Center of the City University of New York, interested in the intersection of computer science education, bilingualism, and social justice pedagogy. Previously, as a senior associate at the youth development organization Global Kids, she developed curricula for and facilitated game design workshops at public schools across NYC, and designed and implemented professional development modules about game design for teachers as part of the NSF-funded iDesign project. She holds a master's degree in bilingual education from CUNY-Hunter College and a BA in urban studies from Columbia University.

Meghan E. Weller is a certified special education teacher and reading specialist in New England with over 7 years of experience. A graduate of Southern Connecticut State University, Meghan holds degrees in special education and remedial reading. She is currently pursuing her doctorate in educational leadership and policy. You can often find her at the bookstore, as she is an avid reader. Meghan lives in Connecticut with her Fiancé Ryan and golden retriever Trigger.

Brent G. Wilson is professor of information and learning technologies at the University of Colorado Denver where he teaches courses in e-learning and instructional design. His research interests center around the following question: What is good instruction? How can we support teachers and instructional designers in creating and delivering good instruction? How can we help people make appropriate use of learning supports and resources? Specific topics of research include professional learning and development; professional practice; integrating social justice and equity concerns into practice; and the role of agency in learning and workplace performance.